NEUROMETHODS

Series Editor
Wolfgang Walz
University of Saskatchewan,
Saskatoon, SK, Canada

For further volumes:
http://www.springer.com/series/7657

Neuromethods publishes cutting-edge methods and protocols in all areas of neuroscience as well as translational neurological and mental research. Each volume in the series offers tested laboratory protocols, step-by-step methods for reproducible lab experiments and addresses methodological controversies and pitfalls in order to aid neuroscientists in experimentation. *Neuromethods* focuses on traditional and emerging topics with wide-ranging implications to brain function, such as electrophysiology, neuroimaging, behavioral analysis, genomics, neurodegeneration, translational research and clinical trials. *Neuromethods* provides investigators and trainees with highly useful compendiums of key strategies and approaches for successful research in animal and human brain function including translational "bench to bedside" approaches to mental and neurological diseases.

More information about this series at http://www.springer.com/series/7657

Cerebrospinal Fluid Biomarkers

Edited by

Charlotte E. Teunissen

Neurochemistry Laboratory, Department of Clinical Chemistry, Amsterdam Neuroscience, Amsterdam University Medical Centre, Vrije Universiteit, Amsterdam, The Netherlands

Henrik Zetterberg

Department of Psychiatry and Neurochemistry, The Sahlgrenska Academy at the University of Gothernburg, Mölndal, Sweden

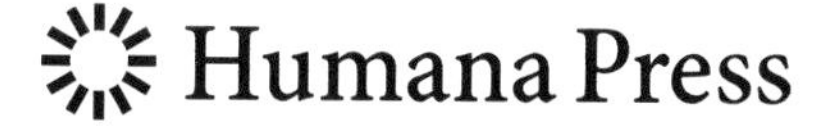

Editors
Charlotte E. Teunissen
Neurochemistry Laboratory
Department of Clinical Chemistry
Amsterdam Neuroscience
Amsterdam University Medical Centre
Vrije Universiteit
Amsterdam, The Netherlands

Henrik Zetterberg
Department of Psychiatry and Neurochemistry
The Sahlgrenska Academy at the University
of Gothernburg
Mölndal, Sweden

ISSN 0893-2336 ISSN 1940-6045 (electronic)
Neuromethods
ISBN 978-1-0716-1321-4 ISBN 978-1-0716-1319-1 (eBook)
https://doi.org/10.1007/978-1-0716-1319-1

This Humana imprint is published by the registered company Springer Science+Business Media, LLC part of Springer Nature.
The registered company address is: 1 New York Plaza, New York, NY 10004, U.S.A.

Series Preface

Experimental life sciences have two basic foundations: concepts and tools. The *Neuromethods* series focuses on the tools and techniques unique to the investigation of the nervous system and excitable cells. It will not, however, shortchange the concept side of things as care has been taken to integrate these tools within the context of the concepts and questions under investigation. In this way, the series is unique in that it not only collects protocols but also includes theoretical background information and critiques which led to the methods and their development. Thus, it gives the reader a better understanding of the origin of the techniques and their potential future development. The *Neuromethods* publishing program strikes a balance between recent and exciting developments like those concerning new animal models of disease, imaging, *in vivo* methods, and more established techniques, including, for example, immunocytochemistry and electrophysiological technologies. New trainees in neurosciences still need a sound footing in these older methods in order to apply a critical approach to their results.

Under the guidance of its founders, Alan Boulton and Glen Baker, the *Neuromethods* series has been a success since its first volume published through Humana Press in 1985. The series continues to flourish through many changes over the years. It is now published under the umbrella of Springer Protocols. While methods involving brain research have changed a lot since the series started, the publishing environment and technology have changed even more radically. Neuromethods has the distinct layout and style of the Springer Protocols program, designed specifically for readability and ease of reference in a laboratory setting.

The careful application of methods is potentially the most important step in the process of scientific inquiry. In the past, new methodologies led the way in developing new disciplines in the biological and medical sciences. For example, Physiology emerged out of Anatomy in the nineteenth century by harnessing new methods based on the newly discovered phenomenon of electricity. Nowadays, the relationships between disciplines and methods are more complex. Methods are now widely shared between disciplines and research areas. New developments in electronic publishing make it possible for scientists that encounter new methods to quickly find sources of information electronically. The design of individual volumes and chapters in this series takes this new access technology into account. Springer protocols make it possible to download single protocols separately. In addition, Springer makes its print-on-demand technology available globally. A print copy can therefore be acquired quickly and for a competitive price anywhere in the world.

Saskatoon, SK, Canada ***Wolfgang Walz***

Preface

The past decade has seen an enormous development of fluid biomarkers for central nervous system diseases such Alzheimer's disease and other neurodegenerative dementias, neuroinflammatory and neuroinfectious conditions. The analysis of cerebrospinal fluid (CSF) is an important aid in the diagnosis of neurological diseases. The number of discriminatory and disease-specific CSF biomarkers is currently increasing, which is the result of intense research, collaboration, and strong technological improvements. Through inter-laboratory comparisons of results from the Alzheimer biomarkers, we have learnt that standardization of laboratory protocols is essential.

The book you hold in your hands presents a reference book to be used in CSF labs and in CSF courses, such as the ones organized by the CSF Society. The target audience is anyone who is involved in clinical neurochemical laboratory work, from undergraduate project students, graduate students, and medical residents to seasoned postdoctoral fellows, research associates, and professors. The book aims to cover all contemporary methods in use in clinical neurochemistry laboratories for both clinical practice and research. Well-tested and basic methodologies are represented in an updated form as their mastery is still very important. These include methods of CSF collection, pre-analytical processing, and basic CSF analysis, such as cell counts and IgG analysis. In addition, the state-of-the-art of analysis of biomarkers is presented. This includes ELISA and more automated immunochemical assays for amyloid and tau markers for Alzheimer's disease. Lastly, there is a special focus on novel cutting-edge techniques. The analysis of neurofilaments by digital ELISA is a typical example, still predominantly used in research settings but in some places now also implemented in clinical laboratory practice. Important technologies for biomarker discovery are discussed, as are important parameters to consider in analytical and clinical biomarker validation. Enjoy!

Amsterdam, The Netherlands — *Charlotte E. Teunissen*
Mölndal, Sweden — *Henrik Zetterberg*

Contents

Contributors

Ulf Andreasson • *Department of Psychiatry and Neurochemistry, The Sahlgrenska Academy at the University of Gothenburg, Mölndal, Sweden; Clinical Neurochemistry Laboratory, Sahlgrenska University Hospital, Mölndal, Sweden*

Christian Barro • *Neurology Clinic and Policlinic, MS Center and Research Center for Clinical Neuroimmunology and Neuroscience Basel (RC2NB), University Hospital Basel, University of Basel, Basel, Switzerland*

Kaj Blennow • *Department of Psychiatry and Neurochemistry, Institute of Neuroscience and Physiology, The Sahlgrenska Academy at University of Gothenburg, Sahlgrenska University Hospital, Mölndal, Sweden; Clinical Neurochemistry Laboratory, Sahlgrenska University Hospital, Mölndal, Sweden*

Nancy Brada • *Department of Pathology and Immunology, Washington University School of Medicine, Saint Louis, MO, USA*

Cyndi Catteau • *CHU de Montpellier, Univ Montpellier, INSERM, Montpellier, France*

Florian Deisenhammer • *Department of Neurology, Neuroimmunology Laboratory, Medical University of Innsbruck, Innsbruck, Austria*

Constance Delaby • *CHU de Montpellier, Univ Montpellier, INSERM, Montpellier, France*

Sebastiaan Engelborghs • *Reference Center for Biological Markers of Dementia (BIODEM), Institute Born-Bunge, University of Antwerp, Antwerp, Belgium; Department of Neurology and Center for Neurosciences, (UZ Brussel) and Vrije Universiteit Brussel (VUB), Brussels, Belgium*

Anne M. Fagan • *Department of Neurology, Knight Alzheimer's Disease Research Center, Hope Center for Neurological Disorders, Washington University School of Medicine, Saint Louis, MO, USA*

Pim Fijneman • *Reference Center for Biological Markers of Dementia (BIODEM), Institute Born-Bunge, University of Antwerp, Antwerp, Belgium*

Audrey Gabelle • *CHU de Montpellier, Univ Montpellier, INSERM, Montpellier, France*

Nelly Ginestet • *CHU de Montpellier, Univ Montpellier, INSERM, Montpellier, France*

Johan Gobom • *Department of Psychiatry and Neurochemistry, Institute of Neuroscience and Physiology, The Sahlgrenska Academy at University of Gothenburg, Mölndal, Sweden*

Alison J. E. Green • *National CJD Research and Surveillance Unit, The University of Edinburgh, Edinburgh, UK*

Svenya Gröbke • *Neurology Clinic and Policlinic, MS Center and Research Center for Clinical Neuroimmunology and Neuroscience Basel (RC2NB), University Hospital Basel, University of Basel, Basel, Switzerland*

Harald Hegen • *Department of Neurology, Neuroimmunology Laboratory, Medical University of Innsbruck, Innsbruck, Austria*

Elizabeth M. Herries • *Departments of Neurology and of Pathology and Immunology, Washington University School of Medicine, Saint Louis, MO, USA*

Christophe Hirtz • *CHU de Montpellier, Univ Montpellier, INSERM, Montpellier, France*

Jens Kuhle • *Neurology Clinic and Policlinic, MS Center and Research Center for Clinical Neuroimmunology and Neuroscience Basel (RC2NB), University Hospital Basel, University of Basel, Basel, Switzerland*

JACK H. LADENSON • *Department of Pathology and Immunology, Washington University School of Medicine, Saint Louis, MO, USA*
SYLVAIN LEHMANN • *CHU de Montpellier, Univ Montpellier, INSERM, Montpellier, France*
PIOTR LEWCZUK • *Lab for Clinical Neurochemistry and Neurochemical Dementia Diagnostics, Department of Psychiatry and Psychotherapy, Universitätsklinikum Erlangen, Erlangen, Germany*
ALEKSANDRA MALESKA MACESKI • *CHU de Montpellier, Univ. Montpellier, INSERM, Montpellier, France*
NEIL I. MCKENZIE • *National CJD Research and Surveillance Unit, The University of Edinburgh, Edinburgh, UK*
ZUZANNA MICHALAK • *Neurology Clinic and Policlinic, MS Center and Research Center for Clinical Neuroimmunology and Neuroscience Basel (RC2NB), University Hospital Basel, University of Basel, Basel, Switzerland*
JOSEF PANNEE • *Department of Psychiatry and Neurochemistry, Institute of Neuroscience and Physiology, The Sahlgrenska Academy at University of Gothenburg, Sahlgrenska University Hospital, Mölndal, Sweden; Clinical Neurochemistry Laboratory, Sahlgrenska University Hospital, Mölndal, Sweden*
AXEL REGENITER • *Medical Laboratories Dr. F. Käppeli, Zurich, Switzerland*
WERNER H. SIEDE • *Medica Medical Laboratories Dr. F. Käppeli, Zurich, Switzerland*
SIGURD D. SÜSSMUTH • *Department of Neurology at RKU, Ulm University Hospital, Ulm, Germany; Boehringer Ingelheim International GmbH, Biberach, Germany*
SARAH STORZ • *Neurology Clinic and Policlinic, MS Center and Research Center for Clinical Neuroimmunology and Neuroscience Basel (RC2NB), University Hospital Basel, University of Basel, Basel, Switzerland*
COURTNEY L. SUTPHEN • *Department of Neurology, Knight Alzheimer's Disease Research Center, Hope Center for Neurological Disorders, Washington University School of Medicine, Saint Louis, MO, USA*
CHARLOTTE E. TEUNISSEN • *Neurochemistry Laboratory, Department of Clinical Chemistry, Amsterdam Neuroscience, Amsterdam University Medical Centre, Vrije Universiteit, Amsterdam, The Netherlands*
HAYRETTIN TUMANI • *Department of Neurology at RKU, Ulm University Hospital, Ulm, Germany*
JÉRÔME VIALARET • *CHU de Montpellier, Univ Montpellier, INSERM, Montpellier, France*
ELINE WILLEMSE • *Neurochemistry Laboratory, Department of Clinical Chemistry, Amsterdam Neuroscience, Amsterdam University Medical Centre, Vrije Universiteit, Amsterdam, The Netherlands*
HENRIK ZETTERBERG • *Department of Psychiatry and Neurochemistry, The Sahlgrenska Academy at the University of Gothernburg, Mölndal, Sweden*

Chapter 1

CSF Cells: Cell Count, Cytomorphology, Cytology, and Immunophenotyping

Sigurd D. Süssmuth and Hayrettin Tumani

Abstract

Assessing the number and type of CSF cells are integral parts of the diagnostic workup of CSF analysis. However, microscopic evaluation of cell morphology provides additional important details that may be missed, if omitted, such as different stages of leukocyte activation, stages of phagocytosis, atypical cells, or tumor cells. While such level of detail may be not required for each and every sample, microscopic evaluation is recommended at least for all cases of differential diagnosis. This chapter provides a review of standard techniques and guidance for interpretation, including comments about potential pitfalls and advice for special cases.

Key words Cerebrospinal fluid, Cells, Leukocytes, Pleocytosis, Microscopic evaluation, Cytomorphology, Cytometry, Immunophenotyping

1 Introduction

The first reported lumbar puncture was performed by Heinrich Irenaeus Quincke in the late nineteenth century in a patient with hydrocephalus [1]. Soon the value of cerebrospinal fluid (CSF) analysis in severe headaches and infections was recognized [2], and despite major advances in diagnostics it is still one of the most important neurological examination methods. Because a lumbar puncture for diagnostic purposes is usually performed only once, as much information as possible should be obtained from the subsequent CSF analysis. The evaluation of cellular components of the CSF are an integral part of the emergency and the basic program of CSF analysis which routinely includes leukocyte and erythrocyte counts, total protein concentration, determination of glucose or lactate, albumin, immunoglobulins (IgG, IgA, and IgM) and assessment of oligoclonal IgG bands. This multiparameter approach allows to reveal acute or chronic inflammatory processes and may lead to further questions for subsequent considerations. In a next step, more specific tests can be performed

Charlotte E. Teunissen and Henrik Zetterberg (eds.), *Cerebrospinal Fluid Biomarkers*, Neuromethods, vol. 168,
https://doi.org/10.1007/978-1-0716-1319-1_1,

depending on the clinical question, e.g., for infections, bleeding, neoplastic diseases, or markers for dementia (extended program). Furthermore, several neuronal and glial biomarkers are frequently used to detect acute brain damage, such as neurofilament protein, protein tau, glial fibrillary acidic protein, and S-100. After analysis of all parameters, an overall assessment of the results should be performed and provided in the form of an integrated cumulative CSF report. This approach enables the detection and allocation of diagnostic patterns which are typical for certain diseases. In addition, this serves as a plausibility control of the individual test results thereby uncovering analytical errors.

The quantitative and qualitative analysis of the CSF cells plays a key role in routine CSF analysis. The leukocyte count is a first important indicator of pathological processes of the central nervous system (CNS). Microscopic evaluation and also immunophenotyping of CSF cells by means of flow cytometry provide essential information about pathological processes in the CSF. Prerequisites for a reliable cell analysis in CSF are careful and quick processing of the CSF sample, reliable cell counting, and microscopic assessment by an experienced cytologist. In order to avoid morphological changes of the cells, cell damage or autolysis, CSF samples should be shipped to the laboratory without any vibration and should be processed within 1 h from collection.

The very first step in CSF analysis is the macroscopic assessment by visual inspection. The CSF is collected into three consecutive sampling containers to allow for a distinction between artificial admixture of blood and states of hemorrhage (three tube test), for example, in subarachnoid hemorrhage (SAH). The degree of redness of the fluid decreases between the individual containers only with artificial blood contamination.

2 CSF Cell Count

2.1 Methods

Traditionally, CSF cells are quantified by microscopic evaluation in native, non-centrifuged samples using the Fuchs-Rosenthal chamber. The chamber volume is 3.2 μL. The counting net consists of 16 × 16 squares to facilitate counting and is limited by an outer triplicate line (Fig. 1.1). Cells touching the outer borderlines are only counted if they touch one of 2 out of the 4 outer edges (e.g., left and bottom lines).

"Cells" always mean all nucleus containing leukocytes irrespective of their subtype. Further differentiation of the uncolored cells is largely impossible by this method. The differentiation of mononuclear cells and polymorphonuclear leukocytes is performed with cytological methods (Subheading 3) or with cytometers. As normal CSF does not contain erythrocytes, these are indicated separately. In patients with known malignancy very large cellular elements may

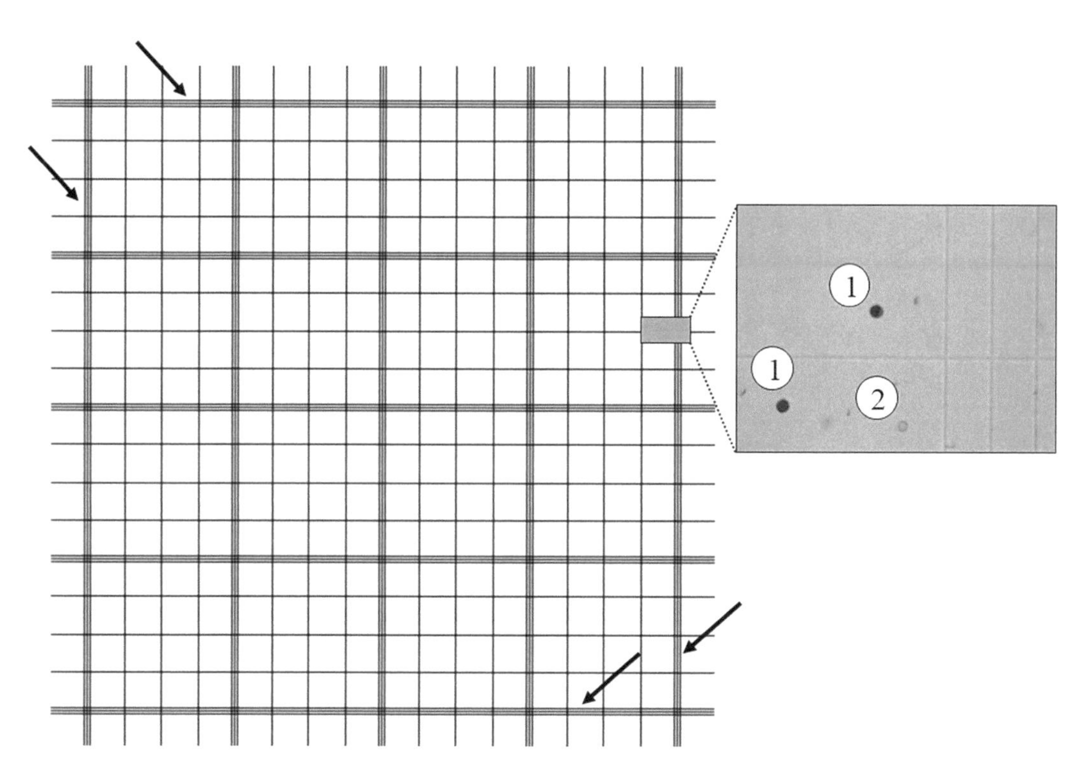

Fig. 1.1 The counting area of the Fuchs-Rosenthal chamber consists of 16 × 16 squares, restricted by the outer triplicate borderlines (→). Leukocytes containing a nucleus (1) and erythrocytes without nucleus but a characteristic ring structure (2) are counted

be an indicator for tumor cells. If cell count is hampered by an overwhelming amount of cells, dilution of the CSF may be required and the dilution factor needs to be taken into account. Leukocytes and erythrocytes are counted and indicated as numbers per μL by using the following calculation:

$$\text{Cell count} \times (1/3.2) \times \text{dilution factor} = \text{cells}/\mu\text{L}.$$

There are companies that offer cytometers for automated counting of CSF cells including cell differentiation. However, these analyzers were primarily designed for counting blood cells. It has been demonstrated that automated cell counting in previous generations of cytometers provided inaccurate results especially at low cell numbers and in case of contamination by blood cells, making the important distinction between normal and slightly increased cell counts unreliable [3, 4]. The analysis of larger volumes of CSF or, in case of blood contamination, dilution may overcome these problems, but in specific cases, a manual count may be still required. Although current generation automated counters do increasingly well, in no case can an automatic cell count replace microscopic differentiation until more multicentre validation studies have been completed [5, 6].

2.2 Interpretation

The normal range for leukocytes in lumbar CSF is <5/μL. An increased leukocyte count is a first important indicator of a pathological process, but conversely, a normal cell count cannot rule out pathological processes [7].

In non-pyogenic meningitis due to viral or atypical bacterial infection, cell counts typically range between 10 and 500 cells /μL with a predominance of mononuclear cells. In some cases, cell counts may reach up to 1000/μL like in some cases of neurotrophic herpes virus infections.

In bacterial meningitis, a huge amount of neutrophilic granulocytes invades into the CSF space and makes the CSF look turbid. Cell counts usually range between 1000/μL and 10–30,000/μL. Following successful antibiotic treatment, a rapid reduction of the cell count is observed.

A number of noninfectious causes may result in low pleocytosis (typically up to 50 cells/μL), like chronic inflammatory diseases, demyelinating polyneuropathy, or ischemic stroke [8].

Erythrocytes may be present in otherwise normal CSF due to artificial blood contamination, or they occur due to hemorrhage into the CSF space.

2.3 Pitfalls

- Air bubbles within the chamber must be avoided because they decrease the total fluid volume, leading to wrong results.
- A vast amount of erythrocytes may mask leukocytes making the cell count unreliable. A dilution of CSF is required.
- In case of bloody CSF, e.g., due to blood contamination or subarachnoid hemorrhage, leukocytes from the blood artificially increase CSF cell counts. A good rule of thumb is to correct down the cell count by one cell per 1000 erythrocytes/μL. This correction should be indicated in the report.
- Neutrophilic granulocytes have a very short half-live in CSF in vitro due to the low protein content and limited buffer capacity of the CSF. Lysis starts even in the first hour after CSF sampling. Thus, granulocyte count is only reliable if the cell count is performed rapidly after lumbar puncture (within the first 2 h at maximum).
- The normal ranges for lumbar CSF cells are age-dependent. CSF white cell counts are higher at birth than in later infancy and fall fairly rapidly in the first 2 weeks of life. In premature infants, the normal cell count is 0–20/μL; in neonates, normal cell count is 0–10/μL; from the third month up to the 15th year, the reference range is 0–5/μL.
- Up to 4 weeks after intrathecal administration of liposomal cytarabine (ara-C) chemotherapy, liposomes may be mistaken for leukocytes by unexperienced raters, leading to incorrect elevated cell counts [9].

3 Cytomorphology

An elevated CSF leukocyte count provides the final diagnosis only very rarely, and conversely, even a normal cell count may be associated with a relevant pathophysiological condition. In case of pathological processes with contact to the CSF space, such as inflammation, bleeding, or tumor cell infiltration, characteristic cytological changes can be found. Therefore, a preparation for differential cell staining should be obtained from each CSF sample for further diagnostic assessment.

Physiologically, the immune cell populations found in normal CSF participate in immune surveillance of the CNS [10]. Figure 1.2 provides an overview about normal CSF cells and examples of cells pointing towards abnormal and pathological processes.

In contrast to the blood, normal CSF contains mononuclear cells only, i.e., nonactivated lymphocytes and monocytes, with strong dominance of lymphocytes between 60 and 90% dependent on the applied preparation method. Ependymal or epithelial cells may rarely be present. Complicated lumbar puncture procedures may be associated with artificial admixture of blood cells, cartilage cells, and occasionally precursor cells from the bone marrow. Exogenous impurities can also occur. Such findings must be recognized and should not be confused, e.g., with tumor cells.

Chronic inflammatory processes and viral inflammations lead to activation of lymphocytes showing an increase in size, expansion of the cytoplasm and maturation up to plasma cells, especially in spirochetal infections such as neuroborreliosis or other non-purulent bacterial infections.

Activation of monocytes is reflected by an enlargement of the cells, formation of intracellular vacuoles, and an increase of cytoplasmic density. Macrophages occur during inflammation or bleeding in order to clear cells, pathogens, or cell debris. They are commonly named after the material they ingest. Thus, leukophages may appear after pleocytosis, lipophages appear in case of parenchymal damage, and erythrophages and siderophages occur during certain disease stages after bleeding (Fig. 1.2). Erythrophages are indicative of hemorrhage and are not present in cases of artificial blood contamination, as phagocytosis occurs approx. four hours after bleeding. Phagocytosis of tumor cells may also occur.

3.1 Special Cases

Pediatric CSF: Apart from different reference ranges regarding the cell count, the lymphocyte/monocyte ratio is shifted towards monocytes. Neonates very often show erythrophages and siderophages although it has not yet been clarified whether such signs of hemorrhage are physiologically caused by the birth trauma or whether an intrauterine bleeding is causal. Physiological CSF of neonates, especially premature infants, may contain immature cells of erythropoiesis and/or granulopoiesis.

Cells in normal CSF	Function	Appearance
Lymphocytes	immunosurveillance	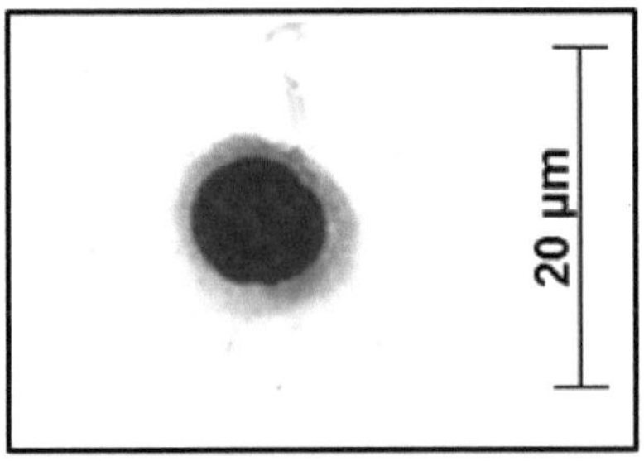 20 µm
Monocytes	first line defence and immunosurveillance	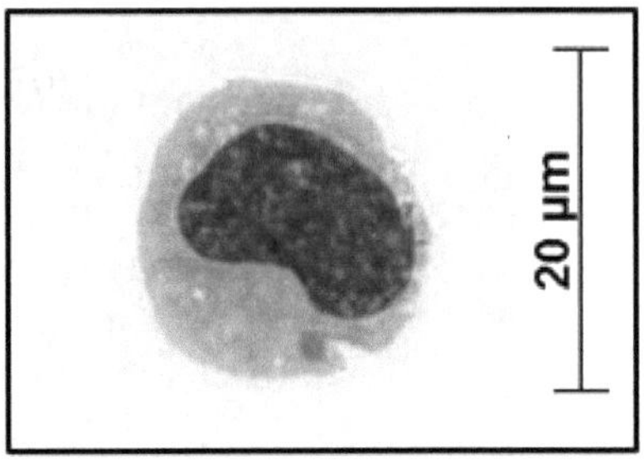 20 µm
Cells indicating a pathological process	**Indication of**	
Activated lymphocyte	acute and chronic inflammation	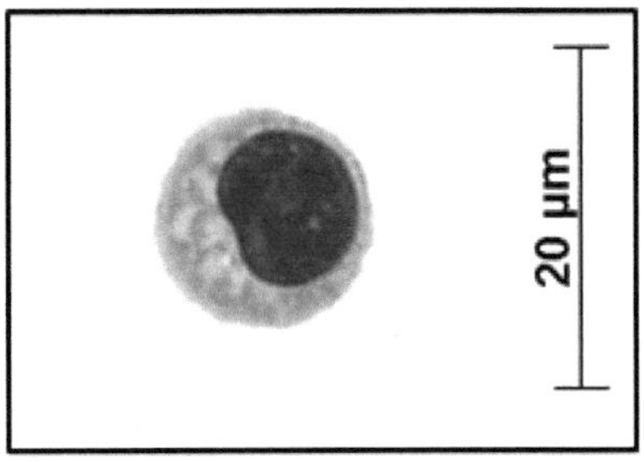 20 µm
Plasma cell	production of antibodies	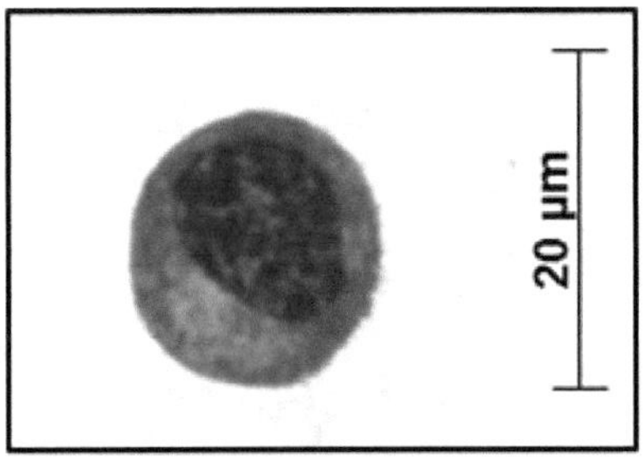 20 µm

Fig. 1.2 Physiological cells and examples of pathological cells in CSF

Macrophage	clearing reaction and recovery phase	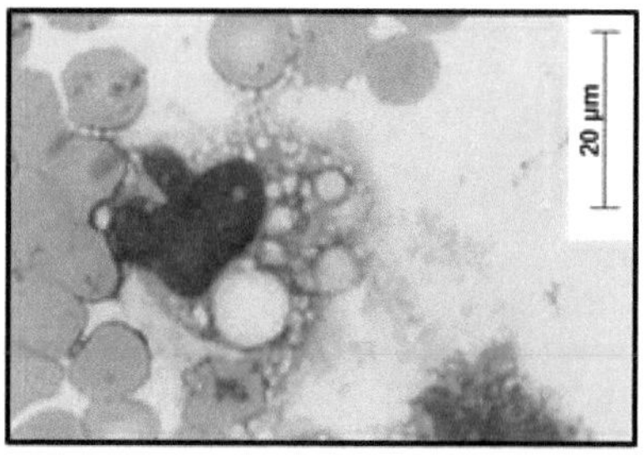
- Leukophage	clearing of cells after pleocytosis	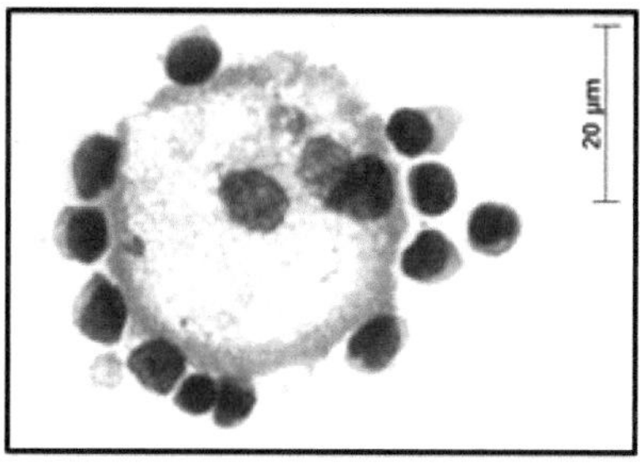
- Erythrophage	haemorrhage >4 hrs after event	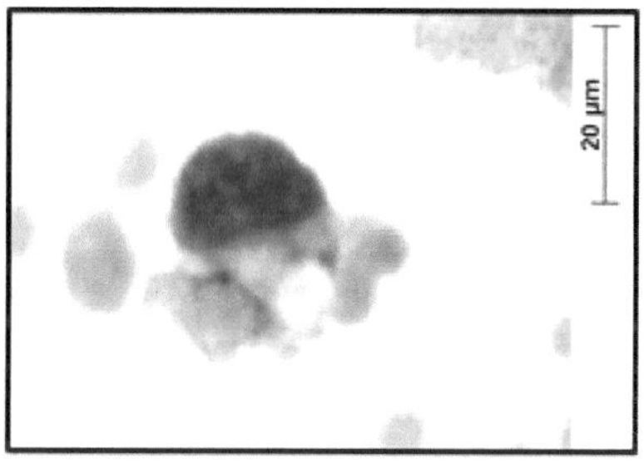
- Siderophage	intracellular hem degradation >10 days after haemorrhage	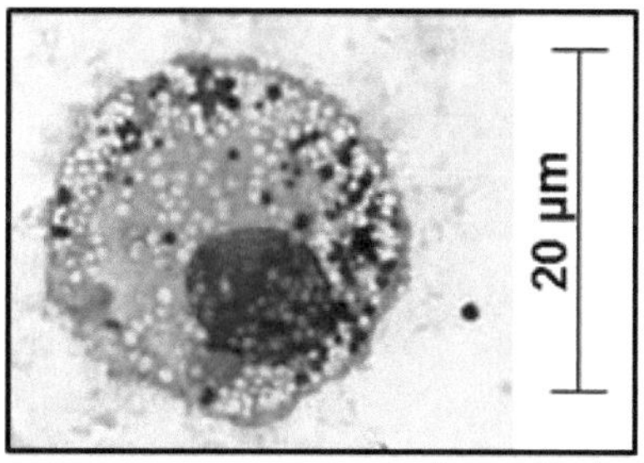

Fig. 1.2 (continued)

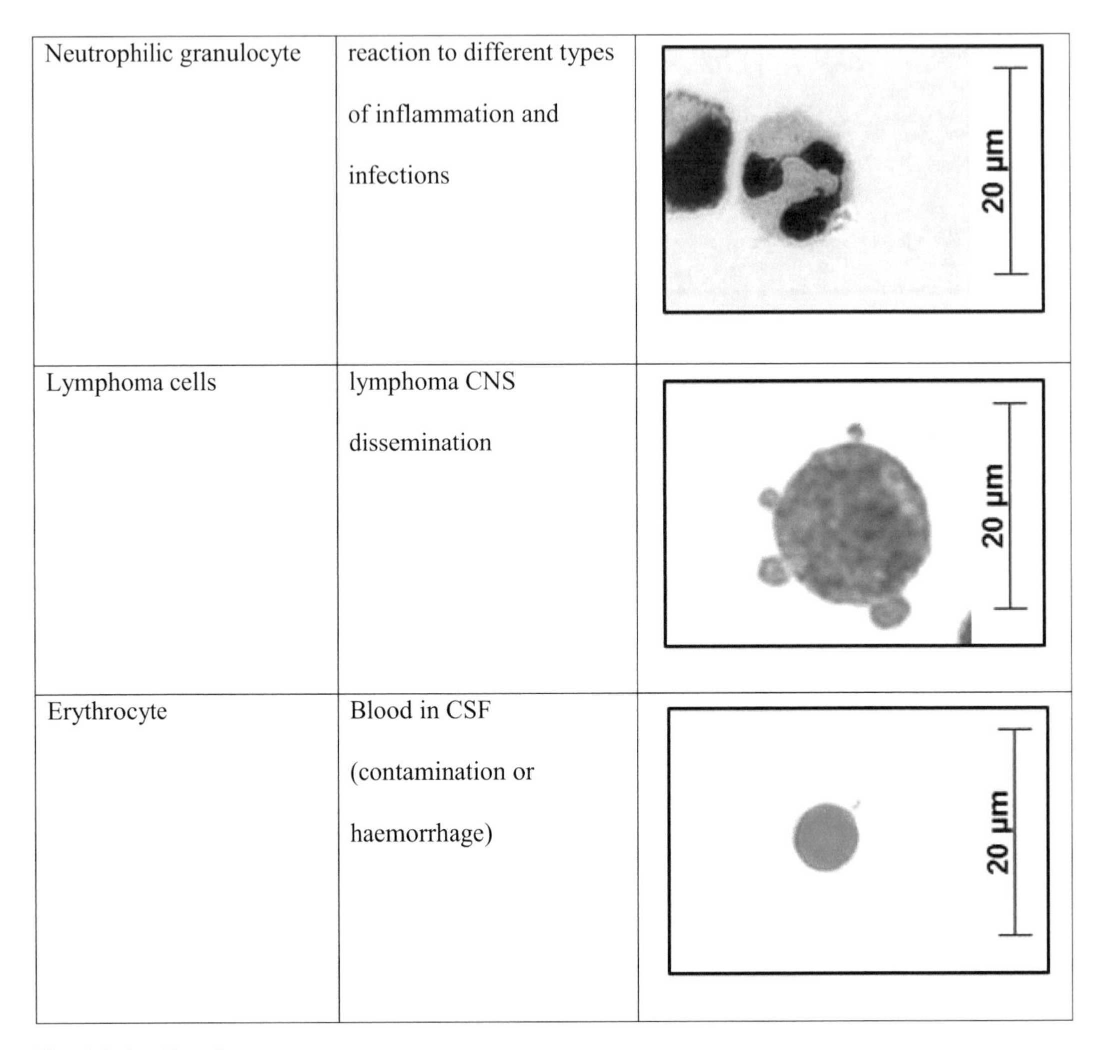

Neutrophilic granulocyte	reaction to different types of inflammation and infections	20 µm
Lymphoma cells	lymphoma CNS dissemination	20 µm
Erythrocyte	Blood in CSF (contamination or haemorrhage)	20 µm

Fig. 1.2 (continued)

Ventricular CSF: Ventricular CSF cell count differs from lumbar CSF by a slightly lower total cell number (<3/µL). Ventricular drainage bears the risk of bacterial immigration. These are often Gram-positive pathogens. Thus, if bacterial infection is suspected, a corresponding second preparation should be made for Gram staining. If different pathogens are found, the sample was most likely contaminated during the collection or preparation procedures. A granulocytic invasion suspicious for acute bacterial meningitis must be assessed in comparison with the hemogram because ventricular samples are often contaminated by blood.

3.2 Methods

Nowadays, cytospin methods are usually used for centrifugation. For cell differentiation, a Pappenheim staining (May-Grünwald-Giemsa) is carried out. Further staining methods are available for specific clinical questions, e.g., Gram for bacteria, Prussian blue (also called Berlin blue) for iron, or Periodic acid–Schiff (PAS) for polysaccharides.

Microscopic evaluation is carried out by an experienced investigator. Both low and high magnification shall be pursued. The detection of a single tumor cell or the presence of erythrophages are of diagnostic relevance.

3.3 Interpretation

While the presence of, e.g., neoplastic cells in CSF often does not require any further CSF tests for confirmation, cytological findings should only be interpreted in combination with other CSF parameters in case of a suspected inflammatory CNS disease (approx. 80% of all questions).

Acute infections usually lead to pleocytosis with invasion of neutrophilic granulocytes into the CSF space. In the very early stage of viral infections, neutrophils predominate, but within a few days mononuclear transformation develops. At the time of lumbar puncture, there is typically already a predominance of mononuclear cells. In contrast, in purulent bacterial meningitis (e.g., by pneumococcus or meningococcus), the leukocytes consist almost exclusively of neutrophilic granulocytes. Following successful antibiotic treatment, reduction in cell counts is associated with mononuclear transformation and a subsequent phagocytic reaction, which is part of the recovery phase. Not every bacterial meningitis is associated with a massive increase in granulocytes, as may be the case in the so-called apurulent bacterial meningitis (mainly just bacteria and minor pleocytosis). With successful therapy, cell count and cytogram often normalize after 2–3 weeks.

Atypical bacterial CNS infections, such as neuroborreliosis or neurosyphilis, have a dominant lymphocytic cytogram with activated lymphocytes and plasma cells. The number of cells varies from several 100/μL to normal cell numbers in the treated late stage. A similar cell pattern can be observed in meningoencephalitis of viral origin. Likewise, only in the early stages abundant granulocytes may occur.

The occurrence of eosinophilic granulocytes typically indicates a reaction to a drug, implants (e.g., shunt), or parasitic infections.

For the detection of SAH and traumatic hemorrhage, cytology has demonstrated its value in addition to cerebral imaging, clinical chemistry (ferritin), and macroscopic assessment of the CSF (three tube sampling test, xanthochromia after centrifugation). It should be done when the quality of the imaging is poor or when blood is not clearly identifiable in the CT scan. A few hours after hemorrhage, erythrophages and activated monocytes occur. After about 3–4 days, hemosiderin appears intra- but also extracellularly as an expression of hemoglobin degradation, which can still be detected in CSF after months. The experienced cytologist can distinguish hemosiderin from melanin in pigmented melanocytes, which also appear as dark granules and may cause confusion. In case of doubt, Prussian blue staining is recommended, which detects the iron in hemosiderin. The final product of hemoglobin degradation is bilirubin, for which the cytologist's term is hematoidin, which occurs extracellularly as shining orange crystals even after months.

The infiltration of the meninges by malignant cells with spread into the CSF is equivalent to a meningeosis neoplastica. In addition to the meningeosis carcinomatosa (usually in the case of breast or bronchial carcinoma), meningeosis lymphomatosa, leucaemica, or sarcomatosa may also occur. In cases with confirmed meningeosis carcinomatosa by detection of malignant cells in the CSF, CSF cell count can be normal [11]. In unselected samples, CSF neoplastic cells are detected in about 1 out of 500, but in patients with hemato-oncological conditions it happens much more often, in about 1 out of 5. Accordingly, careful assessment by experienced investigators is needed. Tumor cells are often significantly different from other cell populations and are eye-catchers when the entire preparation is carefully surveyed. Even in the case of an "empty" counting chamber (0 cells/μL) tumor cells can sometimes be found in the sediment preparation. If, despite clinical suspicion, the detection of tumor cells is not successful in the first examination, a repeated lumbar puncture is advisable with preparation of a cellular sediment from a larger CSF volume. General valid criteria of malignancy are abnormal size and polymorphism of the cells, polychromasia, an increased avidity to the dye, atypical nuclei with abnormal chromatin patterns, multiplication and enlargement of nucleoli, pathological mitosis, amitosis, and a shift of the nucleus/cytoplasm relation in favor of the nucleus (Fig. 1.3). However, their correct allocation requires years of practical experience, and there is sometimes confusion in particular with activated leukocytes. If only few malignancy criteria are met, it is advisable to refer to "tumor suspected" cells.

Overviews of CSF cytology techniques as well as monographies on cytology are listed in the references [12–17].

3.4 Pitfalls

- In case of delayed cell preparation, erythrophages can also be formed due to artificial blood contamination. In such cases, erythrophages do not indicate hemorrhage.
- Artificial blood contamination may lead to transfer from blood granulocytes into the CSF. Especially in the case of inconspicuous further CSF parameters, the mere presence of granulocytes should not be misinterpreted as an expression of acute inflammation.

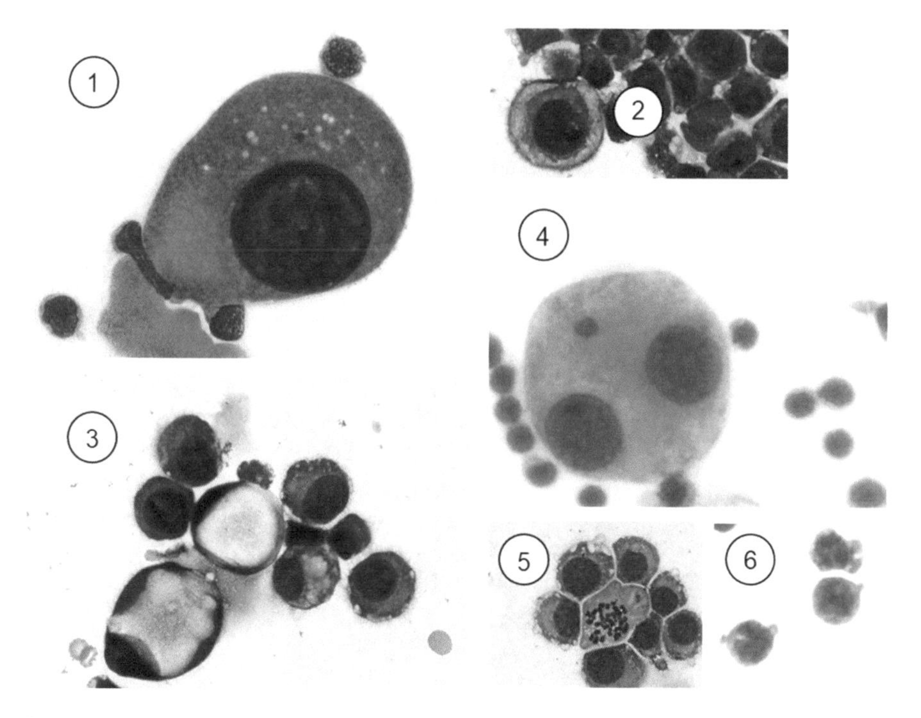

Fig. 1.3 Signs of malignancy. (1) Abnormal large size; increased avidity to the dye; atypic nucleus with abnormal chromatin patterns; (2) polychromasia; (3) cell polymorphism; (4) amitosis; (5) pathological mitosis; (6) shift of the nucleus/cytoplasm relation towards the nucleus. (1–5) carcinomatous meningitis in bronchial carcinoma. (6) CNS lymphoma cells

4 Immunophenotyping of CSF Cells by Flow Cytometry

Immunophenotyping can be performed by immunocytochemistry [18] as well as by flow cytometry [19]. Since flow cytometry is a well-established standardized diagnostic tool in the identification and differentiation of hematological malignancies, immunophenotyping with multi-parameter flow cytometry is increasingly used in the detection of CNS involvement.

4.1 Methods

Flow cytometric immunophenotyping refers to identification and differentiation of various leukocyte subsets and their degree of activation by phenotypic characteristics as revealed by fluorochrome-conjugated monoclonal antibody tags. Polychromatic (multicolor) flow cytometry allows a broad range of immune cells to be phenotyped simultaneously from a single CSF sample. However, the amount of cells can be a limiting factor: at least 1000 cells should be available for each antigen marker. Therefore, cell enrichment is recommended if CSF cell counts are <100 cells/μL; if cell counts are <10 cells/μL, enrichment is an essential prerequisite for reliable results.

Table 1.1
Recommended panel of differentiation markers to determine the composition and the degree of activation of CSF cells, including reference values from normal CSF (adapted from [10])

Marker combination	Subset population	Absolute number[a]
CD45+	Leukocytes (all groups)	1.12 (0.40–3.17)
CD14+	Monocytes/macrophages	0.23 (0.08–1.11)
CD3+ CD56−	T lymphocytes	0.62 (0.15–1.83)
CD3+ CD56− CD4+	T helper cells	0.44 (0.08–1.43)
CD3+ CD56− CD8+	Cytotoxic T cells	0.13 (0.04–0.40)
HLA-DR	Activity marker for T lymphocytes	
CD19+	B lymphocytes	0.005 (0.00–0.03)
CD19+ CD138^{low}	Plasmacytoid cells	0.01 (0.00–0.03)
CD56+ CD3−	Natural killer (NK) cells	0.01 (0.00–0.05)
CD56+ CD3+	Natural killer T cells (NKT)	0.01 (0.00–0.06)
CD16+	Subsets of NK cells and monocytes	

CD cluster of differentiation, *HLA-DR* human leukocyte antigen–DR isotype
[a]Medians (5th–95th percentiles) of absolute numbers/μL are given

4.2 Immunophenotyping of CSF Leukocytes

In accordance with microscopic examinations, the immune profile in the CSF consists of lymphocytes and monocytes [20]. The majority of leukocytes in CSF are T lymphocytes, but compared to blood the CD4/CD8 ratio is increased in CSF. Furthermore, CSF T cells exhibit an increased degree of activation. B lymphocytes and NK cells occur only in small amounts in CSF. It has been shown that the predominating immune cell populations in CSF (e.g., T memory cells, immune regulatory CD56^{bright} NK cells, and plasmacytoid dendritic cells) play a crucial role in immune surveillance [21, 22]. Table 1.1 provides a recommendation for a panel of adequate surface markers for immunophenotyping of CSF leukocytes, together with a compilation of the frequencies of the respective cell types in normal CSF. Studies on the human CSF immunophenotype in healthy individuals and subjects with noninflammatory disorders showed significant differences in the distribution of immune cells in CSF compared to peripheral blood [10, 20].

4.3 Pitfalls

Flow cytometric differentiation of lymphocytes, monocytes, neutrophils, and eosinophilic granulocytes via scatter plots without labeling is very inaccurate. Detection of tumor cells or siderophages is usually even not provided by the software, bearing risks of serious misdiagnosis, such as overlooking a carcinomatous meningitis [3, 4].

4.4 Immunopheno-spi1;typing for Lymphoma in the CNS

Primary CNS lymphomas have increased significantly due to immune suppression treatments and immune deficiency. These are usually highly malignant B-cell non-Hodgkin lymphoma (B-NHL), often associated with Epstein–Barr virus. Initially, they are accompanied by a meningeosis lymphomatosa only in about one third of the cases, with a low proportion of blasts. Also in highly malignant systemic lymphomas (B- and T-NHL), spreading to CNS is common and is usually associated with a meningeosis blastomatosa. Diagnostic problems may be caused by difficulties in distinguishing low-malignant NHL from lymphocytic inflammation. Especially in the rarer occurrences of CNS dissemination in low-malignant lymphomas, only few morphological atypia can make the unambiguous allocation impossible (e.g., chronic lymphocytic leukemia) or, at least, more difficult (e.g., mantle cell lymphoma). If cytomorphological findings are not clearly attributable, immunophenotyping is needed.

In the CSF compartment, aberrant or monoclonal as well as immature cell populations usually do not occur. In normal conditions, hardly any B cells are detectable in the CSF (Table 1.1), and even in inflammatory diseases with strong B-cell stimulation, such as neuroborreliosis, the proportion of B cells is only very rarely >20%. Therefore, predominance of B cells per se is suspicious for a B-NHL. In addition, the identification of a monoclonal population in the sense of an immunoglobulin light chain restriction is a decisive factor for the more frequent B-NHL.

As a minimum program for the detection of B-NHL with very low cell counts, the analysis of CD19/kappa/lambda is required. A major problem can be caused by artificial blood contamination of the CSF so that lysis of the erythrocytes may be required. Moreover, malignant cells from the artificial blood admixture may simulate spreading into the CNS. If possible, a comparative quantitative analysis of blood and CSF may be performed in such cases in order to evaluate the origin of leukemia or lymphoma cells.

5 Clinical Significance of Complementary Cytological Methods

The importance of classical, morphologically based CSF cytology as compared to antibody-based flow cytometric analysis is a matter of debate.

Firstly, the initial situation under which the diagnostics are carried out is essential. In hematological/oncological clinics, the primary neoplasia with its antigens is often already known, so analysis is carried out in a targeted manner. In neurological clinics, the CSF analysis is very often performed without knowledge of the primary neoplasia, so that frequently only an untargeted search is possible.

Comparing the sensitivity of microscopic and flow cytometric approaches, there are very controversial views. A comparison of flow cytometry and conventional cytology in B-cell lymphomas at risk for CNS involvement showed a high sensitivity of flow cytometry compared to a very low sensitivity of cytomorphology, whereas immunocytochemistry was not assessed [23]. If conventional cytology is complemented by immunocytochemistry, the situation is different. Both flow cytometry and immunocytochemistry appear equivalent and reliable with regard to B cell and T cell identification, but immunocytochemistry may be more reliable for examination of low-density surface markers on CSF cells particularly at low cell counts [24]. It is certainly useful to have an integrated analysis that uses both methods to achieve optimal sensitivity and specificity, depending on the specific question [25, 26].

For solid tumors like cancers or melanomas immunocytochemical dyes such as cytokeratins or S-100 could be used in principle. These techniques, however, are poorly standardized and bound to specific laboratories. As a general routine test, they are not widely used due to the rarity of these conditions.

6 Summary

CSF cytology contributes to the diagnosis and differential diagnosis of inflammatory, hemorrhagic and neoplastic processes within or adjacent to the CSF space. Despite modern imaging methods and devices for cell counting and differentiation, conventional cytology continues to be an integral part of CSF analysis and should be carried out without exception at each lumbar puncture, irrespective of the total number of cells. Reliable values for cell counts even at minimum amounts of CSF and at very low cell counts can be obtained by use of the Fuchs-Rosenthal counting chamber. Cell differentiation is based on cytomorphology after Pappenheim staining. If necessary, additional special dyes are used. Cytomorphological assessment requires some experience as well as active feedback interaction with the clinic or the treating doctor. An accurate differentiation of acute, subacute, and chronic inflammatory CNS diseases is often only possible by knowing the cytological pattern. Cytomorphology is supplemented by modern immunocytochemical and flow cytometric techniques, in particular for immunophenotyping of cells in inflammatory CNS diseases and for the differential diagnosis in lymphomas.

Acknowledgments

We thank all technicians, scientists, and physicians of the CSF laboratory of the Department of Neurology at the University Hospital Ulm. Our special thanks go to Dagmar Vogel, Dr. Andreas Junker, MD, and Dr. Vera Lehmensiek, PhD for preparation of the cyto-images.

References

1. Quincke H (1891) Die Lumbalpunktion des Hydrocephalus. Berl Klin Wochenschr 28:929–933
2. Lundie A, Thomas DJ, Fleming S (1915) Cerebro-spinal meningitis: diagnosis and prophylaxis is lumbar puncture justifiable? Br Med J 1:628–629
3. Heller T, Nagel I, Ehrlich B et al (2008) Automated cerebrospinal fluid cytology. Anal Quant Cytol Histol 30:139–144
4. Strik H, Luthe H, Nagel I et al (2005) Automated cerebrospinal fluid cytology: limitations and reasonable applications. Anal Quant Cytol Histol 27:167–173
5. Bremell D, Mattsson N, Wallin F et al (2014) Automated cerebrospinal fluid cell count—new reference ranges and evaluation of its clinical use in central nervous system infections. Clin Biochem 47(1–2):25–30
6. Hod EA, Brugnara C, Pilichowska M et al (2018) Automated cell counts on CSF samples: a multicenter performance evaluation of the GloCyte system. Int J Lab Hematol 40 (1):56–65
7. Kölmel HW (2003) Konventionelle Liquorzytologie. In: Zettl U, Lehmitz R, Mix E (eds). Klinische Liquordiagnostik. De Gruyter, Berlin. isbn:3-11-016846-4, pp 135–160
8. Østergaard AA, Sydenham TV, Nybo M, Andersen AB (2017) Cerebrospinal fluid pleocytosis level as a diagnostic predictor? A cross-sectional study. BMC Clin Pathol 17:15
9. Strik H (2015) Cell count and staining. In: Deisenhammer F, Sellebjerg F, Teunissen CE, Tumani H (eds) Cerebrospinal fluid in clinical neurology. Springer, Basel., isbn:978-3-319-01225-4, pp 81–100
10. de Graaf MT, de Jongste AHC, Kraan J et al (2011) Flow cytometric characterization of cerebrospinal fluid cells. Cytometry B Clin Cytom 80:271–281
11. Djukic M, Trimmel R, Nagel I, Spreer A, Lange P, Stadelmann C, Nau R (2017) Cerebrospinal fluid abnormalities in meningeosis neoplastica: a retrospective 12-year analysis. Fluids Barriers CNS 14(1):7
12. Den Hartog Jager WA (1980) Color atlas of Csf cytopathology. Lippincott Williams & Wilkins, Philadelphia
13. Fishman RA (1992) Cerebrospinal fluid in diseases of the nervous system. W.B. Saunders, Philadelphia
14. Kölmel HW (1986) Zytologie des Liquor cerebrospinalis. Edition Medizin, VCH, Weinheim
15. Oehmichen M (1976) Cerebrospinal fluid cytology: An introduction and atlas. W.B. Saunders, Philadelphia
16. Schmidt RM (1978) Atlas der Liquorzytologie. Barth, Leipzig
17. Worofka B, Lassmann J, Bauer K et al (1997) Praktische Liquorzelldiagnostik. Springer, Wien - New York
18. Kranz BR (1991) Methodik und Wert immunzytochemischer Differenzierung benigner und maligner Zellen im Liquor cerebrospinalis. Lab Med 15:61–68
19. Kleine TO, Albrecht J (1991) Vereinfachte Druchflusszytometrie von Liquorzellen mit FACScan. Lab Med 15:73–78
20. Lueg G, Gross CC, Lohmann H et al (2015) Clinical relevance of specific T-cell activation in the blood and cerebrospinal fluid of patients with mild Alzheimer's disease. Neurobiol Aging 36:81–89
21. Gross CC, Schulte-Mecklenbeck A, Runzi A et al (2016) Impaired NK-mediated regulation of T-cell activity in multiple sclerosis is reconstituted by IL-2 receptor modulation. Proc Natl Acad Sci U S A 113:E2973–E2982
22. Han S, Lin YC, Wu T et al (2014) Comprehensive immunophenotyping of cerebrospinal fluid cells in patients with neuroimmunological diseases. J Immunol 192:2551–2563
23. Hegde U, Filie A, Little RF et al (2005) High incidence of occult leptomeningeal disease detected by flow cytometry in newly diagnosed aggressive B-cell lymphomas at risk for central

nervous system involvement: the role of flow cytometry versus cytology. Blood 105:496–502

24. Windhagen A, Maniak S, Heidenreich F (1999) Analysis of cerebrospinal fluid cells by flow cytometry and immunocytochemistry in inflammatory central nervous system diseases: comparison of low- and high-density cell surface antigen expression. Diagn Cytopathol 21:313–318
25. Bommer M, Nagy A, Schopflin C et al (2011) Cerebrospinal fluid pleocytosis: pitfalls and benefits of combined analysis using cytomorphology and flow cytometry. Cancer Cytopathol 119:20–26
26. Sancho JM, Orfao A, Quijano S et al (2010) Clinical significance of occult cerebrospinal fluid involvement assessed by flow cytometry in non-Hodgkin's lymphoma patients at high risk of central nervous system disease in the rituximab era. Eur J Haematol 85:321–328

Chapter 2

Nephelometry and Turbidimetry: Methods to Quantify Albumin and Immunoglobulins Concentrations in Clinical Neurochemistry

Piotr Lewczuk

Abstract

CSF concentrations of albumin and immunoglobulins (IgG, IgA, and IgM) are contemporary analyzed by fully automated, particle-enhanced turbidimetry or nephelometry. Turbidimetry measures decrease of the intensity of light passing through a solution (a CSF sample) due to its scattering on the particles suspended in this sample. Nephelometry applies a similar principle, but with a light detector placed at an angel to the source. This allows measurement of the intensity of the light reflected on the particles suspended in a sample. Diagnostically relevant are concentrations of albumin, IgG, IgA, and IgM. To normalize their CSF concentrations for the concentrations in the blood, the results are expressed as CSF/serum concentrations quotients (i.e., Q_{Alb}).

Since albumin is generated exclusively in the liver, it diffuses passively into the CSF through the blood–CSF barrier, and it is not metabolized by the brain, Q_{Alb} is the biomarker of choice for the blood–CSF barrier function. In contrast, under pathologic conditions, like neuroinflammations, immunoglobulins may be synthesized intrathecally. Two phenomena need be considered to correctly interpret immunoglobulins concentrations in the CSF: (a) passive diffusion through the blood–CSF barrier and (b) possible release from intrathecal lymphocytes. Hence, "references" for the CSF immunoglobulins quotients exist only dependent on the blood–CSF barrier staus (i.e., Q_{Alb}).

Hyperbolic functions proposed by H. Reiber in 1990s describe the correlation between Q_{Alb} and Q_{Ig}, enabling correct interpretation of the immunoglobulins concentrations found in the CSF. Furthermore, distinctive features of the intrathecal immune response, characteristic for the CSF, need to be taken into consideration to properly interpret albumin/immunoglobulins patters.

Key words Turbidimetry, Nephelometry, Blood–CSF barrier function, Reibergram, Neuroinflammation

1 Introduction

In modern clinical neurochemistry, the concentrations of albumin and the three diagnostically relevant immunoglobulins (IgG, IgA, and IgM) are measured exclusively by fully automated particle-enhanced turbidimetry or nephelometry. Other methods are solely of historical or theoretical meaning.

Charlotte E. Teunissen and Henrik Zetterberg (eds.), *Cerebrospinal Fluid Biomarkers*, Neuromethods, vol. 168, https://doi.org/10.1007/978-1-0716-1319-1_2,

Turbidimetry measures decrease of the intensity of light passing through a solution (a sample) due to its scattering on the particles suspended in this sample. In clinical neurochemistry, to enhance analytical sensitivity latex particles coated with antibodies against albumin, IgG, IgA, or IgM are incubated with a solution to form the antigen–antibody complexes, causing turbidity of the sample and loss of the intensity of the light beam passing through it. Turbidity (loss of the light's intensity) is proportional to the concentration of the analyte.

Nephelometry applies a similar principle, but with a light detector placed at an angel (e.g., 90°) to the source. In contrast to turbidimetry, the intensity of the light is measured reflected on the particles (antigen–antibody complexes) suspended in a sample. In case of IgA and IgM, latex particles are used to enhance analytical sensitivity of the method. Intensity of the reflected light is proportional to the concentration of the analyzed protein.

2 Albumin Quotient (Q_{Alb}): A Biomakrer of the Blood–CSF Barrier Function

In contrast to the blood–brain barrier, which has a well-defined morphologic substrate (brain endothelial cells), the blood–CSF barrier is not anatomically localized. Instead, it is defined as the mechanism responsible for passive diffusion of the blood proteins into the CSF. It is worth stressing that ***all*** proteins existing in the blood are also found in the CSF; their CSF concentrations are only the matter of the functional status of the blood–CSF barrier, a given protein's molecular size (which can be approximated to some extent by the molecular weight), and the presence/absence of its intrathecally synthesized fraction.

The sole source of albumin in human blood is the liver. Albumin's blood concentration is very high, it enters the CSF exclusively by passive diffusion through the blood–CSF barrier, and it is not metabolized in the brain. These characteristics make albumin an ideal biomarker of the blood–CSF barrier function. A concentration of albumin in the CSF is influenced by its concentrations in the blood, which can fluctuate physiologically, but it can also be altered in diseases not relevant for clinical neurochemistry (an obvious example is decreased blood albumin concentration in liver disorders). Therefore, instead of using raw CSF albumin concentration as a biomarker of the blood–CSF barrier function, it is more plausible to normalize it for its concentration in the blood. By measuring the albumin concentrations in the CSF and in the serum (or plasma) with the same method, in the same analytical run (and of course on the same analyzer), and with the same calibration curve, albumin CSF/serum concentration quotient (Q_{Alb}) can be calculated as:

$$Q_{\mathrm{Alb}} = \frac{\text{Albumin in CSF}}{\text{Albumin in serum}}.$$

For practical reasons, to avoid fractions, this quotient (and similar quotients discussed in the next paragraph) is conventionally multiplied by 1000. Hence, for example, $Q_{\mathrm{Alb}} = 6.0$ means that the albumin's concentration in the CSF equals to 0.6% of its concentration in the serum or, equivalently, that the blood–CSFconcentration gradient is approximately 167:1.

Blood–CSF barrier function depends on age, and hence to correctly interpret Q_{Alb} it is crucial to apply age-dependent reference values. The physiologic function of the barrier (or, more exactly spoken, the CSF flow rate) in newborns and young babies is not fully mature; in newborns the physiological Q_{Alb} may well exceed 20, a value strongly pathologic in adults. Due to a rapid maturation of the CSF flow, a reference for Q_{Alb} decreases to reach the lowest values of 3.5 at the age of 6 months, and then it steadily increases. This age-dependent upper limit of Q_{Alb} can be calculated for the subjects between the age of 5 and 60 years according to the formula [1]:

$$Q_{\mathrm{Alb}} = 4 + \frac{\text{Age [years]}}{15}$$

For persons older than 60, the upper limit for Q_{Alb} remains 8.0 (in some laboratories it is extended to 9.0 for patients between 60 and 75).

CSF does not circulate. It is generated in a more or less well-defined part of the brain, it flows in the subarachnoid space, and it is absorbed back to the blood, or it is collected by a puncture. A given "portion" of the CSF never returns to its place of origin, which means that in contrast to the blood samples, the composition of the CSF samples strongly depends on the anatomical localization of the CSF withdrawal. The CSF concentrations of the blood-derived proteins, like albumin and immunoglobulins, increase along the spinal cord. This phenomenon, which is referred to as a *rostro-caudal concentration gradient*, is easily explainable: the more time a given portion of the CSF has on its flow from its origin (brain ventricles) to its destination (a test tube), the more molecules are diffusing from the blood into the CSF, following their concentration gradient. Note however, that the gradient-dependent diffusion of the brain-derived proteins from the CSF into the blood is much more complex and still not fully understood [2]). For albumin (and Q_{Alb}), the rostro-caudal gradient is about 1:2.5. Therefore, to calculate a reference value for Q_{Alb} in the CSF collected by a ventricular puncture (VP), the above-presented age-dependent Q_{Alb} limit for the lumbar CSF has to be multiplied by 0.4. For example, in case of a 27-year-old patient, the reference values of

Q_{Alb} are 5.8 and 2.3 in the lumbar and the ventricular CSF, respectively.

Finally, it is worth mentioning that the mechanism responsible for the functional status of the blood–CSF barrier is the CSF flow rate.

3 IgG, IgA, and IgM Quotients ($Q_{IgG/IgA/IgM}$). Intrathecal Fractions of the Blood-Derived Proteins. Hyperbolic Functions (Reibergrams)

In contrast to albumin, which enters the CSF exclusively by a passive diffusion from the blood, many other predominantly blood-borne proteins may be in addition synthesized intrathecally. It means that they may have, under particular physiologic or pathologic conditions, an *intrathecal fraction* (IF). This fraction can be estimated, as a percentage of the total concentration of a given protein X in the CSF, according to [3]:

$$IF_X = \left(1 - \frac{Q_{XLim}}{Q_X}\right) * 100\%$$

where Q_{XLim} is the upper limit of the quotient of a given protein X in the CSF, derived from the appropriate hyperbolic function (see below this paragraph), and Q_X is its actually measured quotient.

Although the immunoglobulins quotients are calculated the same way as the Q_{Alb} (concentration in the CSF divided by the concentration in the serum and multiplied by 1000 for convenience), their interpretation is entirely different, as two phenomena need be considered: (a) passive diffusion through the blood–CSF barrier and (b) possible release from lymphocytes migrating into the central nervous system under certain pathologic condition, like neuroinflammation. As a consequence, "references" for the CSF immunoglobulins concentrations, and even for the immunoglobulins quotients, do not exist as numerical values, but are dependent on the status of the blood–CSF barrier. For example, a 10-times increased Q_{IgG} may simply result from a passive diffusion of this immunoglobulin from the blood (in case of a severe dysfunction of the blood–CSF barrier, for example in Guillain-Barré-Syndrom, GBS), or it may be brought about by active intrathecal synthesis of the immunoglobulin (e.g., in neuroinflammation, like MS), or may be a net effect of both processes. If increased Q_{IgG} results solely from the diffusion from the blood, it is interpretationally meaningless; however, if it is, at least partly, caused by intrathecal synthesis, it is extremely relevant in differential diagnostics. Therefore, an approach to answer the question on the existence/absence of an IF of a blood-borne protein in the CSF has to consider the (dys)function of the blood–CSF barrier, as reflected by the age-dependent Q_{Alb} and the relation of the diffusion rate of these two molecules through the barrier.

Many linear and nonlinear formulas correlating Q_{Alb} and Q_{IgG} (or Q_{IgA} or Q_{IgM}) were considered in the past, all of them aiming to answer the same question, if a given concentration of the immunoglobulin in the CSF can be fully attributed to its passive diffusion from the blood, or if some of its portion is released intrathecally. After heavy and sometimes very hot discussions in the 1980s and 1990s, it became clear that the *hyperbolic functions* proposed by H. Reiber not only fit the empirical data but can be directly derived from the Fick's laws of diffusion. Unfortunately, step-by-step deriving of the Reiber's formulas exceeds the scope and the volume of this book, so the interested readers are kindly referred to the excellent publications explaining it further: [3–5]. The general formula for a hyperbolic function, correlating a quotient of a given immunoglobulin (Q_X) and Q_{Alb} is:

$$Q_X = \frac{a}{b} * \sqrt{Q_{Alb}^2 + b^2} - c$$

where a, b, and c are parameters of the function, depending on the molecular size of the immunoglobulin in question (note that to correctly estimate these three parameters, Q_X and Q_{Alb} need to be expressed without the 1000-multiplication factor). Further, it also follows from the hyperbolic model of the molecular size-dependent protein diffusion that $Q_{Alb} > Q_{IgG} > Q_{IgA} > Q_{IgM}$. Violation of this dependence (e.g., $Q_{IgG} \geq Q_{Alb}$) is an evidence for intrathecal synthesis of a given immunoglobulin irrespectively of the actual values of the quotients.

In the next step, a graphic representation of the hyperbolic functions were developed depicting the $Q_{Alb}/Q_{IgG/A/M}$ hyperbolic reference curves, which tremendously simplified the interpretation. As a respectful appreciation of the Hansotto Reiber's input to the field, his students and coworkers (but not he himself) began calling these graphs the "*Reibergrams,*" and this expression was promptly adopted by the research community. A Reibergram for IgG and its interpretation is presented in Fig. 2.1.

4 A Couple of Examples of the Patterns of the CSF/Serum Quotients

Detailed discussion of clinical neuroimmunology, even restricted to its diagnostic aspects, would exceed the scope of this book. Nevertheless, some distinctive features, characteristic for the CSF, need to be briefly mentioned to understand the principles of the interpretation of the humoral immune reaction in the CSF [3]:

(a) "IgM/IgG switch," typically observed in the blood in the course of infectious disorders, is usually not observed in the

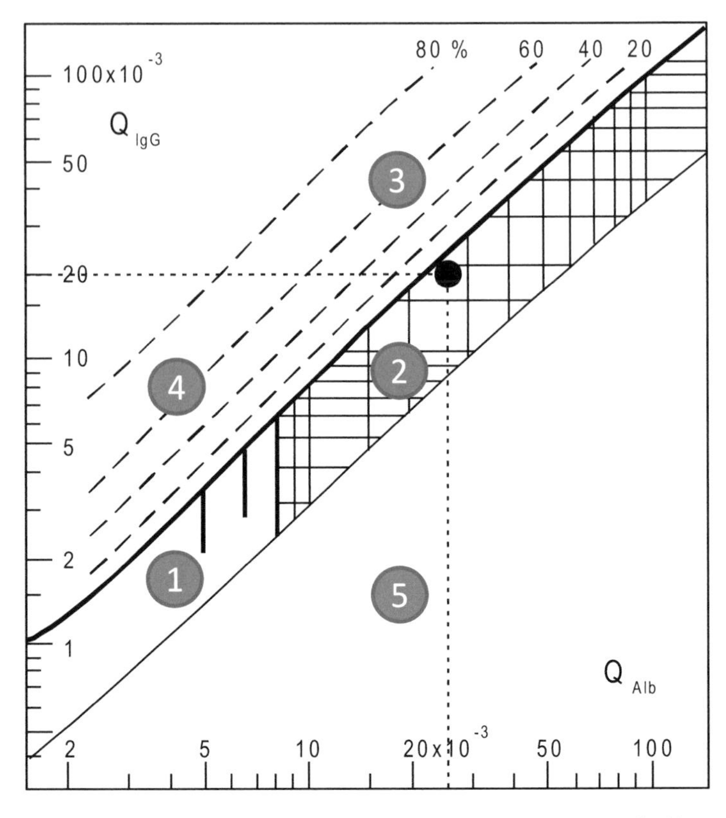

Fig. 2.1 Reibergram for IgG. The solid curves represent the upper (bold curve) and the lower limits of Q_{IgG} as a hyperbolic function of Q_{Alb}. Dashed curves can be used to roughly estimate intrathecal fraction of the immunoglobulin as percentage of its total concentration in the CSF. Three solid vertical lines between the two limit-curves represent Q_{Alb} reference value for a 15, 45, and 60-year-old subjects, respectively. The five numbered circles represent the five areas of the Reibergram, to be interpreted as: (1) normal blood–CSF function and no intrathecal IgG synthesis; (2) dysfunction of the barrier without intrathecal IgG synthesis; (3) dysfunction of the barrier and intrathecal IgG synthesis; (4) normal blood–CSF barrier and intrathecal IgG synthesis; (5) area of analytical implausibility (the whole area below the lower limit curve). Note logarithmic scales of the axes. A circle on the intersection of the two dotted lines represents the results ($Q_{Alb} = 24$, $Q_{IgG} = 20$) obtained in patient, which should be interpreted as *dysfunction of the blood–CSF barrier without intrathecal IgG synthesis*. Note that ~5-times increased CSF IgG concentration and, correspondingly, Q_{IgG}, as compared to a "normal" case, is interpretationally irrelevant in this subject, and should be treated as normal, since it results from a passive diffusion of the immunoglobulin through the dysfunctional blood–CSF barrier. (Reproduced with slight modifications from www.horeiber.de)

CSF. As a consequence, the dominance of the intrathecal IgM synthesis should not be interpreted as a sign of acuity of a neuroinfection. A good example is neuroborreliosis, a disease with persisting domination of the IgM synthesis.

(b) Very often a poly-specific, oligoclonal immune response is observed in the CSF with a fraction of the antibodies specific for a causative antigen around 30% or below 1% in case of stimulation of the antibodies against other antigens (discussed in more details in the chapter on the antibody indices and the MRZH-Reaction).

(c) Persistence, or very slow decrease, of intrathecal immune synthesis may be observed years or even decades after successful treatment (the so-called *immune scar*).

Taken together, a similar pattern of the blood–CSF barrier dysfunction/intrathecal immunoglobulin synthesis can be observed in diseases with entirely different pathophysiology. For example, domination of the IgM intrathecal synthesis, irrespectively of the stage of neuroinfection, is almost always observed in neuroborreliosis (Fig. 2.2).

Figure 2.3 presents CSF/serum quotients patterns in multiple sclerosis, bacterial meningitis, and neurotuberculosis. A typical pattern in MS (Fig. 2.3a) shows a normal or only slightly abnormal blood–CSF function with dominating intrathecal IgG synthesis, sometimes accompanied by intrathecal synthesis of other immunoglobulins. Such pattern, very often with a slight blood–CSF barrier dysfunction and without IgM or IgA synthesis, is also seen in viral meningoencephalitis. Bacterial meningitis (Fig. 2.3b) characterizes with a very profound blood–CSF barrier dysfunction, which rapidly normalizes under successful antibiotic treatment, and without intrathecal immunoglobulin synthesis, at least during the first days of the disease. The same pattern can be observed in GBS and in disk prolapse. Neurotuberculosis (Fig. 2.3c) is marked with very pronounced blood–CSF barrier dysfunction and dominating IgA synthesis. A similar pattern, with IgA dominance but usually with much lower Q_{Alb} is frequently observed in abscesses of the central nervous system.

5 Tips, Difficulties, and Pitfalls

1. Decreasing concentrations with increasing collected CSF volume; blood contamination

   The rostro-caudal gradient of the blood-derived proteins in the CSF has important practical consequence: by withdrawal of more than 10 mL of the fluid, which should be a normal case in the clinical routine to assure the appropriate CSF volume for all analyses, the decrease in the protein concentrations between the first and the last milliliter can exceed 20%. Therefore, it is strongly recommended to collect all the CSF volume into one test tube, and to revert the tube gently once or twice after the

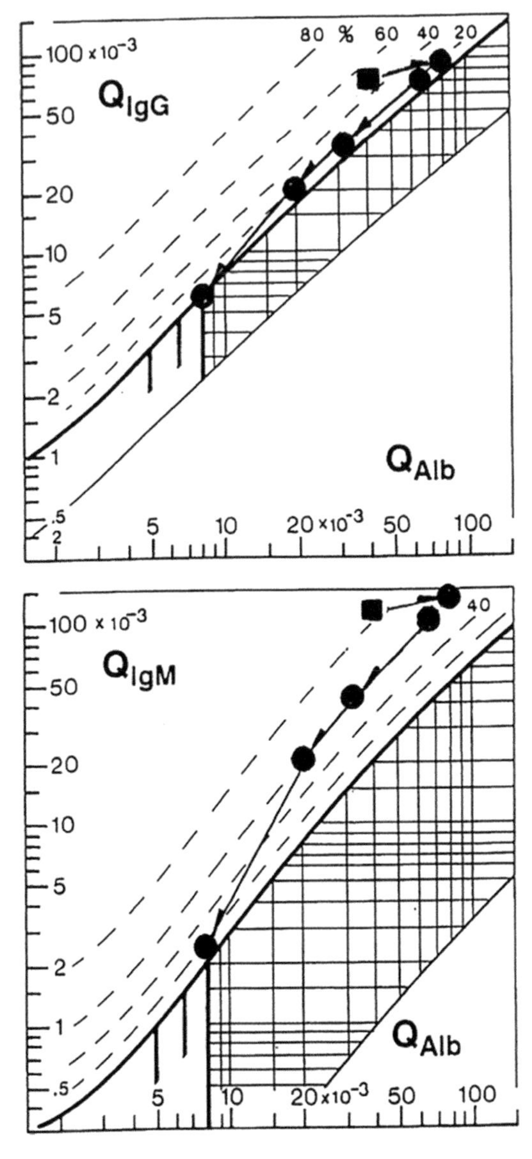

Fig. 2.2 A pattern of the blood–CSF barrier function and IgG (upper) and IgM (lower) quotients in the follow-up of a case with clinically definite neuroborreliosis. The first diagnostic LP marked with a square. Note persistent dominance of intrathecal IgM synthesis during the whole course of the disease, and a steady normalization of the blood–CSF barrier function. (Reprinted from: J Neurol Sci (2001), vol. 184, Reiber H. and Peter J. B. Cerebrospinal fluid analysis: disease related data patterns and evaluation programs, pages 101–122; Copyright (2001), with kind permission from Elsevier)

LP to homogenize the sample (do not forget to close the tube!).

In case of blood contamination (see also the "three-tubes-test"), very restricted corrections of the CSF leukocytes count,

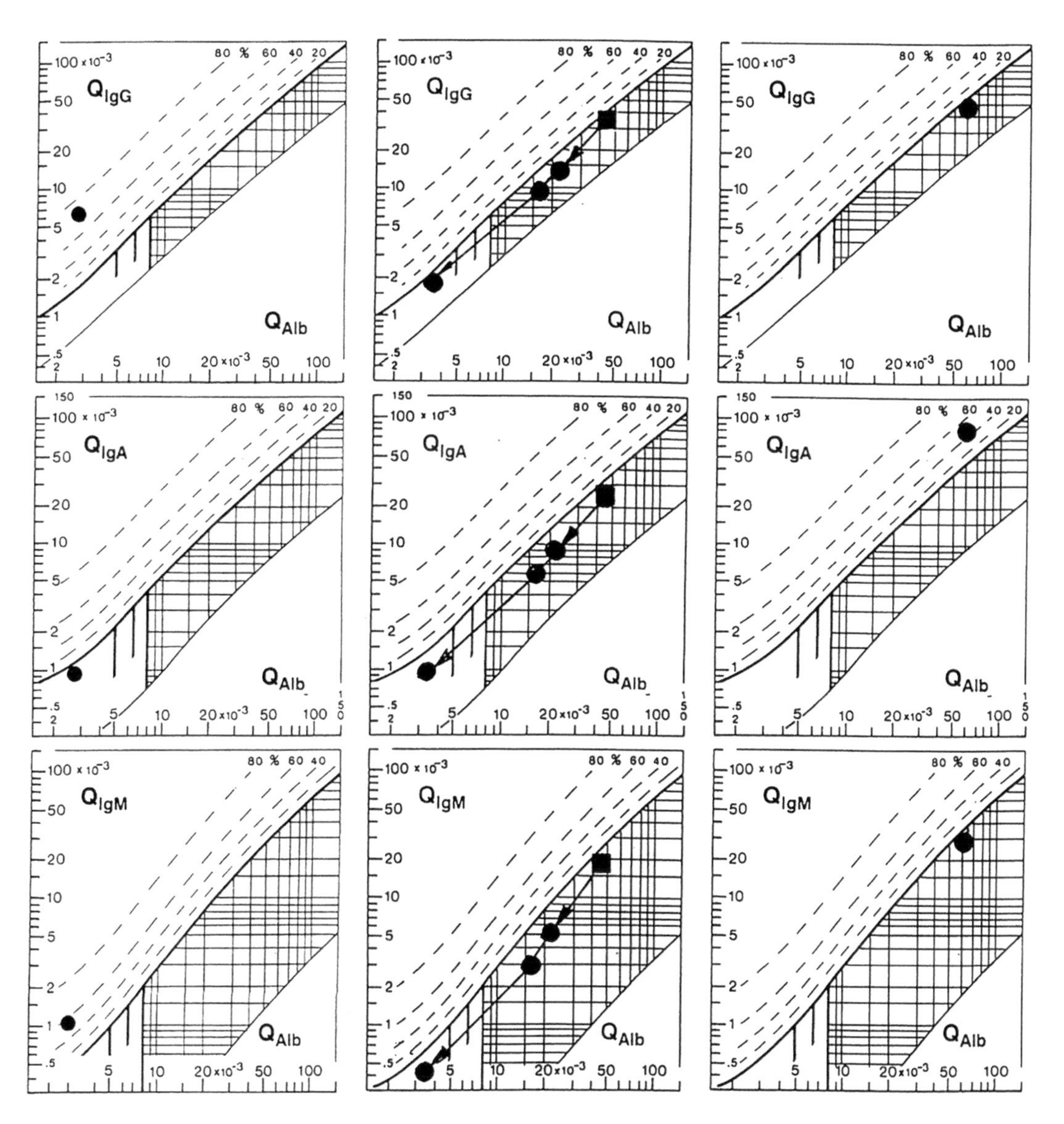

Fig. 2.3 Exemplary patterns of the Reibergrams in: (**a**) multiple sclerosis; (**b**) bacterial meningitis (follow-up LPs); (**c**) neurotuberculosis. (Reprinted from: J Neurol Sci (2001), vol. 184, Reiber H. and Peter J. B. Cerebrospinal fluid analysis: disease related data patterns and evaluation programs, pages 101–122; Copyright (2001), with kind permission from Elsevier)

and albumin and IgG concentration might be applied if the erythrocytes count does not exceed 7000/μL. Such corrections should be based on the actual values measured in the simultaneously collected blood sample. In principle, also here the rule $Q_{Alb} > Q_{IgG} > Q_{IgA} > Q_{IgM}$ is valid, but in case of severe blood contamination, the points corresponding to the albumin and Ig's quotients will be approaching a 45°-line on the Reibergrams, which may lead to false-positive interpretation, particularly in case of Q_{IgM} and when Q_{Alb} is low [3].

According to the author's opinion, results from a CSF sample containing more than 7000 erythrocytes per one μL should not be interpreted.

2. Simultaneous collection of the blood and CSF samples. Steady-state conditions of protein diffusion

   The CSF/serum quotients reflect the correlation between the blood–CSF barrier function and the intrathecal Ig's synthesis only if the two body fluids are collected simultaneously and under the steady-state diffusion conditions. "Simultaneous" means that the CSF and the blood samples should be withdrawn within a time frame not exceeding 30 min, irrespectively what is collected first.

   In addition, the issue of the steady-state condition of the proteins diffusion is particularly relevant in such cases, when the concentrations of the proteins of interest rapidly change in the blood, for example, due to infusions or plasmapheresis. Albumin concentrations reach blood–CSF steady-state in approximately 1–2 days, much bigger IgG needs up to 4 days. Ignoring it leads to false interpretation of the results, with potentially severe diagnostic consequences. For example, in patients with GBS treated with plasmapheresis, falsely positive intrathecal synthesis of immunoglobulins (very high $Q_{IgG/IgA/IgM}$) is observed when the treatment is introduced before the puncture [6]. This interpretational pitfall can be avoided by observing heavily decreased concentrations of immunoglobulins in the blood, which explains that very high CSF/serum immunoglobulins quotients in these cases result from the decreased denominators (concentrations in the blood) and not increased numerators (concentration in the CSF) of the quotients. Further, the intrathecal IgG synthesis is never confirmed by isoelectrofocusing in these subjects (almost always type 4 is observed; see the chapter on IEF).

3. Use the same method, calibrator, and the analytical run!

   Application of the CSF/serum quotients, instead of the raw concentrations, of the blood-derived proteins has an additional advantage: systematic measurement errors of the method compensate. This means that if the systematic measurement error equals, say, +5%, it will affect the concentrations of a given protein in both samples (the numerator and the denominator of the quotient) to the same extent. But this advantage exists only when the protein in question is analyzed in both CSF and serum samples with the same method, on the same analyzer, with the same reagents, and in the same analytical run. Otherwise, the errors may even propagate if, by chance, the systematic errors of the CSF- and the serum- methods drive the results in the opposite directions. An economic counterargument (due to much lower protein concentrations in the CSF, the methods to analyze them need to be more sensitive, and hence they are more expensive) should not counterbalance the advantages of the improvement in the quality of the analyses. Therefore, it is strongly recommended that the prediluted serum samples are analyzed together with the CSF samples and with the "CSF reagents," and not separately with the "Serum reagents."

4. Discrepancies between Reibergrams and IEF. Presence of an IF when the Reibergram is "negative."

   An integrated, comprehensive CSF laboratory report is an excellent tool also in a sense of plausibility of the results interpretation. In the absolute majority of cases, the presence/absence of the IF of IgG on an IgG-Reibergram corresponds to the presence/absence of IgG IF on an isoelectrofocusing (IEF). However, IEF is analytically more sensitive than nephelometry, and hence it captures intrathecal fractions also when the Reibergrams are negative. This is also a consequence of the fact that turbidimetry and nephelometry are quantitative methods and IEF is a qualitative one. Hence, a relatively small but diagnostically relevant IF of IgG would contribute to the total net CSF IgG not necessarily to such extent that Q_{IgG} would exceed the upper reference limit. Such IF would, however, be visible as additional CSF band(s) on the IEF. In everyday practice, cases are observed with borderline negative IgG-Reibergrams (a point just below the curve) but with clear intrathecal synthesis on the IEF. In contrast, extremely rare are the vice versa cases, with positive IgG-Reibergram but without presence of the additional CSF IgG bands on the IEF. So, a practical advice here is that if the IEF is positive (type 2 or 3) and the IgG-Reibergram negative (particularly, borderline negative)—that is probably analytically fine; however, if the IgG-Reibergram is positive and the IEF negative (type 1 or 4), it is worth to repeat the analyses.

References

1. Reiber H, Otto M, Trendelenburg C, Wormek A (2001) Reporting cerebrospinal fluid data: knowledge base and interpretation software. Clin Chem Lab Med 39:324–332
2. Brandner S, Thaler C, Lelental N, Buchfelder M, Kleindienst A, Maler JM, Kornhuber J, Lewczuk P (2014) Ventricular and lumbar cerebrospinal fluid concentrations of Alzheimer's disease biomarkers in patients with normal pressure hydrocephalus and posttraumatic hydrocephalus. J Alzheimers Dis 41:1057–1062
3. Reiber H, Peter JB (2001) Cerebrospinal fluid analysis: disease related data patterns and evaluation programs. J Neurol Sci 184:101–122
4. Reiber H (1994) Flow rate of cerebrospinal fluid (CSF)—a concept common to normal blood-CSF barrier function and to dysfunction in neurological diseases. J Neurol Sci 122:189–203
5. Reiber H (1995) Biophysics of protein diffusion from blood into CSF: the modulation by CSF flow rate. In: Greenwood J (ed) New concepts of a blood-brain barrier. Plenum Press, New York, pp 219–227
6. Madzar D, Maihofner C, Zimmermann R, Schwab S, Kornhuber J, Lewczuk P (2011) Cerebrospinal fluid under non-steady state condition caused by plasmapheresis. J Neural Transm 118:219–222

Chapter 3

Oligoclonal Bands: Isoelectric Focusing and Immunoblotting, and Determination of κ Free Light Chains in the Cerebrospinal Fluid

Harald Hegen and Florian Deisenhammer

Abstract

A variety of inflammatory diseases of the central nervous system ranging from an autoimmune to infectious pathophysiology are characterized by intrathecal B cell activity. Immunoglobulins (Ig) and free light chains (FLC) which are both secreted by terminally differentiated B cells can be detected in the cerebrospinal fluid (CSF). As these proteins are not only present in the CSF in case of intrathecal inflammation, but also derive from blood by diffusion across the blood–CSF barrier, methods for the detection of an intrathecal synthesis needs to take this into account.

For the determination of an intrathecal IgG synthesis, isoelectric focusing followed by immunoblotting is the gold standard. This technique reveals oligoclonal bands (OCB) and depends on comparing paired CSF and blood samples of each individual patient. An intrathecal IgG synthesis is present if OCB are detected in CSF without corresponding bands in serum. For the determination of an intrathecal κ-FLC synthesis, first κ-FLC concentrations are measured in CSF and serum, usually by nephelometry using antibodies against κ-FLC-specific epitopes, and then referred to an upper normal limit. For this comparison, different approaches can be applied, among others the calculation of the κ-FLC index.

In this chapter, we provide background information on OCB and κ-FLC, describe the technology used for their determination, discuss their applications in clinical practice and how to interpret the obtained results.

Key words Cerebrospinal fluid, Oligoclonal bands, IgG, Isoelectric focusing, Immunoblotting, Kappa free light chains, Nephelometry

1 Oligoclonal Bands

1.1 Introduction

Intrathecal immunoglobulin (Ig) synthesis occurs in different inflammatory diseases of the central nervous system (CNS) ranging from an autoimmune to infectious pathophysiology [1]. Its determination is part of routine cerebrospinal fluid (CSF) workup [2] and supports clinical diagnoses such as multiple sclerosis [3].

Immunoglobulins are secreted by terminally differentiated B cells and can be measured in the cerebrospinal fluid (CSF). Immunoglobulins in CSF originate either from blood by diffusion

Charlotte E. Teunissen and Henrik Zetterberg (eds.), *Cerebrospinal Fluid Biomarkers*, Neuromethods, vol. 168,
https://doi.org/10.1007/978-1-0716-1319-1_3,

across the blood–CSF barrier or are locally synthesized within the CNS [4]. For diagnostic purposes, determination of the locally synthesized fraction of immunoglobulins, which is distinct from the blood-derived fraction, is essential.

For the detection of an intrathecal IgG production, different laboratory methods are available. Besides quantitative methods that require the measurement of IgG concentrations in CSF and serum followed by calculations of certain formulae such as IgG index [5], Reiber [6], or Auer & Hegen formulae [7] referring patient's individual values to a predefined upper normal limit, isoelectric focusing (IEF) followed by immunoblotting has evolved as gold standard method [8]. This technique reveals oligoclonal IgG bands (OCB) and depends on comparing paired CSF and blood samples of each individual patient. An intrathecal IgG synthesis is present, if OCB are present in CSF without corresponding bands in serum. Due to this direct comparison and the fact that this method enables only qualitative determination of an intrathecal IgG synthesis (i.e., positive or negative result), there is no need to correct for blood–CSF barrier function in contrast to the above-mentioned formulae.

In the following, we first provide the background on IgG as a biomarker, describe the technology used for the detection of OCB, display the different OCB patterns that might be obtained, and discuss the application and value of OCB in clinical practice.

1.1.1 Immunoglobulin G

IgG and all other immunoglobulin isotypes are secreted by terminally differentiated B cells. They share the same basic structural characteristics, as they are composed of two identical heavy chains and two identical light chains. Each heavy chain consists of four immunoglobulin domains linked by a hinge region. Differences in the structure of the constant regions (C_H1, C_H2, and C_H3) determine the isotype (IgG, M, A, D, E) and subclass of the immunoglobulin (e.g., IgG1-4). The amino-terminal variable domain (V_H) participates in antigen recognition. Each light chain consists of two immunoglobulin domains and has a molecular weight of approximately 24 kD. Differences in the structure of the constant region (C_L) determine the isotype of free light chain (either κ or λ). The variable domain (V_L) contributes to the antigen-binding site. Accordingly, both the heavy chains and light chains consist of amino-terminal variable (V) regions responsible for antigen recognition and carboxyl-terminal constant (C) regions that mediate effector functions (Fig. 3.1) [9]. When finally assembled, IgG reaches a total molecular weight of approximately 150 kD.

1.1.2 History of Intrathecal IgG Detection

A century ago, the colloidal gold test, a precipitation method which specifically detects globulins excluding albumin, was introduced by Carl Friedrich Lange [10]. Further progress was achieved by Elvin Kabat and colleagues, who introduced electrophoresis technique in clinical neurology and demonstrated an increase of

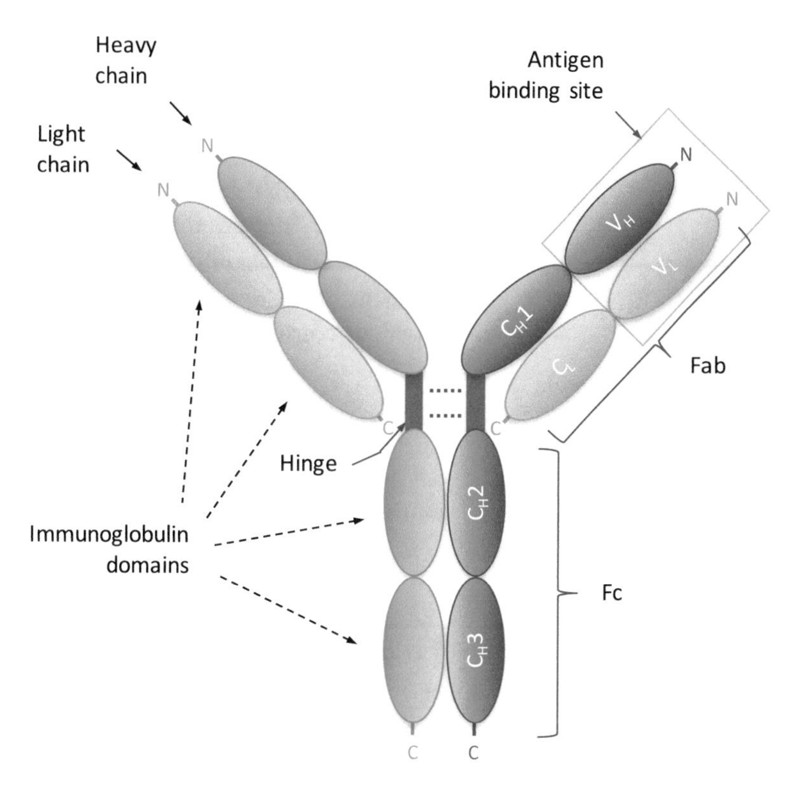

Fig. 3.1 Schematic presentation of the basic structure of immunoglobulins. C_H constant heavy chain domain, C_L constant light chain domain, *Fab* fragment antibody binding, *Fc* fragment crystallizable, V_H variable heavy chain domain, V_L variable light chain domain

immunoglobulins in the CSF independent of serum concentrations, particularly in patients with multiple sclerosis and neurolues [11]. These results corroborated the prior hypothesis that the origin of antibodies in CSF is two-fold: "from the blood and from the cerebrospinal tissue" [12]. The refinement of electrophoresis toward higher resolution of immunoglobulins came in the late 1950s driven by Ewald Frick [13] and Hans Link. The use of agar gel as supporting medium led to sharp bands that represented immunoglobulins of different classes besides free light chains and other CNS-derived proteins [14]. Delmotte eventually introduced IEF to demonstrate CSF-restricted oligoclonal bands [15]. The final significant step forward was achieved by combining highly sensitive IEF on agarose gel with subsequent immunoblotting to detect IgG-specific OCB by immunostaining methods [16]. Until today, this procedure is still widely used in diagnostic CSF workup with some methodological variations.

1.1.3 Principles of Isoelectric Focusing

Isoelectric focusing is an electrophoretic procedure that separates proteins according to their isoelectric point. Following simplified principles can be considered as the basis for this method [17]:

1. Proteins are ampholytes, which possess both basic and acidic functional groups within their structure (especially amino and carboxyl residues) and, thus, may act either as an acid or a base.
2. Whether a protein acts as acid or base depends on the number and type of ionisable chemical groups as well as the pH of the environment, determining the overall charge of the protein. This relationship is illustrated by the following examples:
   - Carboxyl residues release a proton in basic environment, which contributes to a negative overall charge of the protein:

   $$R - COOH + OH^- \Leftrightarrow R - COO^- + H_2O$$

   - Carboxyl residues receive a protein in acidic environment and get uncharged subsequently:

   $$R - COO^- + H^+ \Leftrightarrow R\text{–} COOH$$

   - Amino residues take up a proton in acidic environment contributing to a positive overall charge of the protein:

   $$R - NH_2 + H^+ \Leftrightarrow R\text{–} NH_3^+$$

   - Amino residues release a proton in basic environment and get uncharged subsequently:

   $$R - NH_3^+ + OH^- \Leftrightarrow R\text{–} NH_2 + H_2O$$

3. The environmental pH at which a protein possesses no overall charge is defined as its isoelectric point. As the primary structure differs between proteins (different amino acid sequences), proteins show different isoelectric points. With regard to IgG, amino acid sequences are different between the subclasses IgG1, IgG2, IgG3, and IgG4. Furthermore, there are differences in the light chains (κ or λ), in the variable region (there are $>10^9$ distinct IgG molecules differing in their variable region) as well as in the posttranslational modifications. These differences do not significantly impact the total molecular weight of IgG, but its isoelectric points. That is why IEF and not common electrophoresis methods is able to separate different IgG clones.
4. If a protein is not located at its isoelectric point, it will be charged (either positive or negative). When subjected to an electric field within gel electrophoresis, the protein will move toward the electrode of opposite charge.
5. If a pH gradient is established within the gel between the electrodes, the charged chemical groups of the protein are neutralized while migrating through the gel by the increasing concentration of oppositely charged ions (as derived from the environment). At a certain point, the protein possesses no

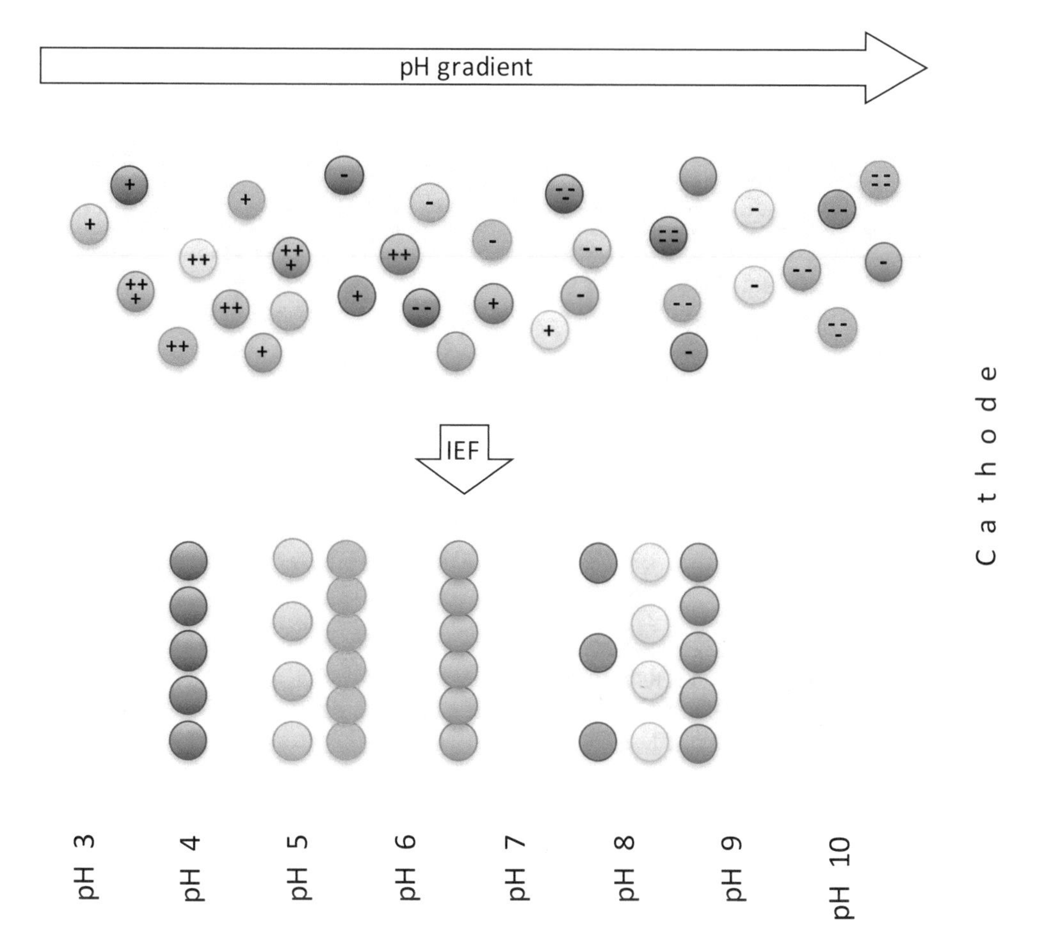

Fig. 3.2 Principles of isoelectric focusing. If proteins are not at their isoelectric point, they will by definition possess an overall charge. If subjected to a pH gradient and an electric field within gel electrophoresis, proteins start migrating toward the electrode of opposite charge and stop when their overall charge is zero, i.e., when they have reached their isoelectric point. Therefore, similar proteins stop at the same location and can be focused into tight bands

overall charge, therefore, stops migration in the electric field (Fig. 3.2). IgG are typically clonally restricted in case of intrathecal production; therefore, there are "enough" similar IgG molecules that stop migration in the electric field at the same location, i.e., they can be focused into a tight band.

1.1.4 Interpretation of IEF

Patterns of Oligoclonal IgG Bands

IEF is performed using paired CSF and serum samples. A classification of OCB into five different patterns has been proposed by two consensus reports and is nowadays widely accepted (Fig. 3.3) [8, 18]:

Type I: No bands in CSF and serum

This "normal" pattern reflects the presence of polyclonal IgG in both CSF and serum; there is no evidence of clonally restricted B cell activity.

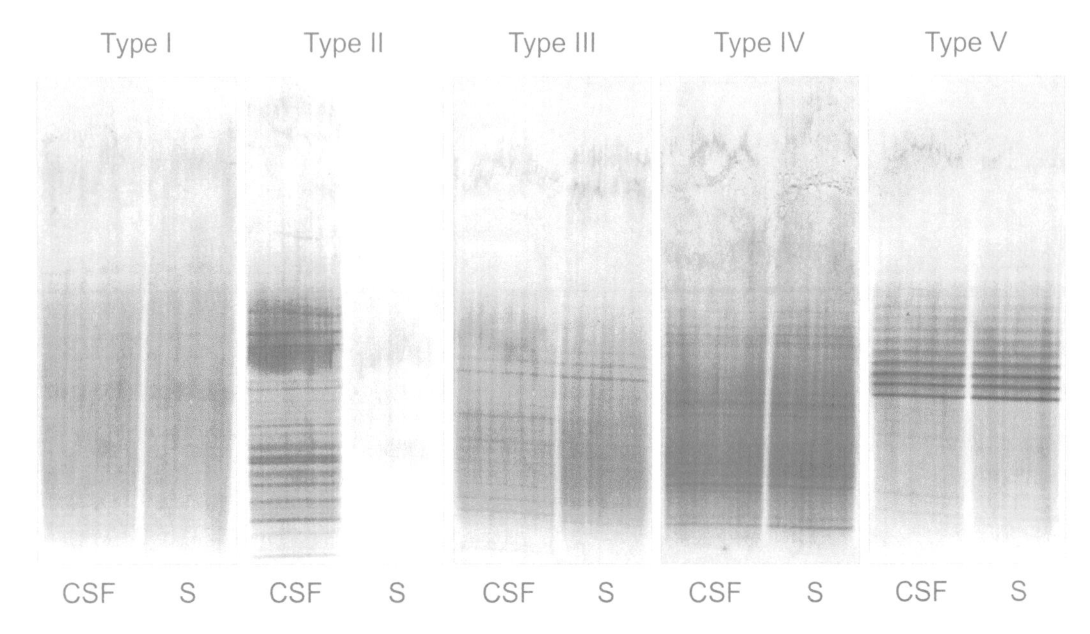

Fig. 3.3 Patterns of oligoclonal IgG bands. Type I, No bands in CSF and serum. Type II, OCB present in CSF without corresponding bands in serum. Type III, OCB present in both CSF and serum, with additional bands present in CSF. Type IV, OCB present in CSF which are identical to those in serum. Type V, OCB present in both CSF and serum showing a "ladder type"

Type II: OCB present in CSF without corresponding bands in serum

This pattern comprises bands present in CSF while absent in serum indicating an intrathecal synthesis of IgG.

Type III: OCB present in both CSF and serum, with additional bands present in CSF

This pattern shows a mixture of pattern type II and type IV. It comprises OCB in the CSF without corresponding bands in serum, superimposed to bands present in both CSF and serum. This result indicates an intrathecal IgG synthesis in addition to a systemic immune response.

Type IV: OCB present in CSF which are identical to those in serum

This pattern shows for each CSF band a corresponding band in serum ("mirror pattern"), i.e., OCB show identical migration (position) and occur at identical numbers in both CSF and serum. This is the result of passive transfer of IgG from serum into CSF due to a systemic immune (B cell) response, e.g., in case of systemic infection.

Type V: OCB present in both CSF and serum showing a "ladder type"

This pattern is characterized by several regularly spaced bands and indicates the presence of monoclonal gammopathy (M protein). The source of these monoclonal IgG's lies outside the CNS typically occurring in patients with monoclonal gammopathy of undetermined significance (MGUS) or myeloma. The fact that IEF resolves what would be a single band using other electrophoretic techniques into multiple bands might be explained by posttranslational modifications such as glycosylation of monoclonal IgGs.

Definition of OCB Positivity

The number of OCB present in CSF but absent in serum to define an intrathecal IgG synthesis (pattern II or III) is not well established. Some studies—all of them determining OCB by IEF and subsequent immunoblotting, either on agarose or polyacrylamide gel—applied the definition of >1 band, whereas others applied the definition of >2 bands in order to assess diagnostic accuracy. In multiple sclerosis—the most frequently studied disease with regard to OCB—the diagnostic sensitivity reaches approximately 95%, similarly whether >1 [19, 20] or >2 OCB [21] was used as cut-off. Studies including healthy subjects or patients with no obvious neurological disease such as primary headache, neck pain, or lower back pain calculated a diagnostic specificity of approximately 95% using a cut-off >1 OCB [22] and up to 100% using a cut-off >2 OCB [21, 23]. Besides that, there is some evidence that the presence of one OCB in CSF but not in serum has no reliable clinical significance [24].

Taken together, there is no clear evidence whether a cut-off >1 or >2 OCB is better to decide if intrathecal IgG synthesis is present or not. There is only one study that compared patients with MS to patients with other neurological diseases and reported superior diagnostic accuracy of a cut-off >2 OCB than >1 OCB for MS, however, missed to explicitly describe the method of IgG staining [25]. As detection of OCB is of main relevance in the diagnosis of MS and as McDonalds criteria allow to make the diagnosis based on only one clinical attack and additional paraclinical findings including positive CSF OCB [3], a high diagnostic specificity is preferred over diagnostic sensitivity.

1.1.5 Application in Clinical Practice

The detection of OCB, i.e., of clonally restricted, intrathecally synthesized IgG, is a hallmark of multiple sclerosis [26]. OCB are observed in approximately 95% of MS patients [19–21]. OCB detection is part of the MS diagnostic criteria and can substitute dissemination in time otherwise shown by clinical disease activity or magnetic resonance imaging metrics [3]. However, OCB are not specific to multiple sclerosis or any other neurological disease. OCB provide evidence for chronic intrathecal immune activity and can be found in a variety of inflammatory CNS disorders besides multiple sclerosis, such as systemic lupus erythematosus affecting the CNS, neurosarcoidosis, Behcet's disease, vasculitis, and different infectious diseases including viral or bacterial meningitis, neuroborreliosis, and neurosyphilis (Table 3.1) [2]. Identification of antigens that are targeted by IgG OCB revealed that in CNS infections OCB contain antibodies directed against the etiologic agent such as VZV [30] whereas no consistent results could be obtained in disorders such as MS [31].

Detection of OCB is technically demanding, time consuming, and rater-dependent [32]. There are other quantitative methods available for the detection of intrathecal IgG synthesis, e.g., calculation of the intrathecal fraction by Reiber formula after

Table 3.1
Incidence of OCB in different inflammatory CNS diseases

Disease	OCB positive
Demyelinating-inflammatory	
Multiple sclerosis	95%
Neuromyelitis optica	<20%
Acute disseminated encephalomyelitis	<15%
Autoimmune	
Neuro-SLE	50%
Neuro-Behcet's	20%
Neuro-sarcoidosis	40%
Harada's meningitis-uveitis	60%
Infectious	
Acute viral encephalitis (<7 days)	<5%
Acute bacterial meningitis (<7 days)	<5%
Subacute sclerosing panencephalitis	100%
Progressive rubella panencephalitis	100%
Neurosyphilis	95%
Neuro-AIDS	80%
Neuroborreliosis	80%
Hereditary	
Ataxia teleangiectasia	60%
Adrenoleukodystrophy	100%

AIDS acquired immunodeficiency syndrome, *CNS* central nervous system, *OCB* oligoclonal band, *SLE* systemic lupus erythematodes
Adapted after [2] and complemented by [27–29]

nephelometric measurement of IgG concentrations in CSF and serum. However, detection of OCB by IEF and immunoblotting shows a considerably higher sensitivity [20, 21, 33] and, thus, is still considered as gold standard method.

1.2 Materials

The technique of IEF might slightly differ between laboratories. This variability includes, e.g., different gels such as agarose or polyacrylamide gel, different sample application, different sample volume at different IgG concentration, different parameters for carrying out IEF (in terms of applied voltage or duration) or different detection methods (labeling of the detection antibody with, e.g., alkaline phosphatase or horseradish peroxidase [HRP]). Here, we describe IEF using polyacrylamide gel as previously published [34].

1.2.1 Test Reagents

In the following, all buffers, reagents and solutions are listed for production of a 7.5% polyacrylamide gel and for carrying out IEF as previously described [34].

- Acrylamide solution: 6 g acrylamide and 0.12 g bis-acrylamide diluted in 20 mL distilled water (final acrylamide concentration 30%).
- Ammonium persulfate: 0.12 g ammonium persulfate diluted in 10 ml distilled water
- Carrier ampholyte: covering a pH range of 3–10
- Blocking solution: 3 g dried, skimmed milk dissolved in 150 mL 0.9% NaCl
- Antibody diluent: 20 mL blocking solution diluted in 200 mL NaCl
- Substrate solution: 25 mg 3-amino-9-ethylcarbazole, 10 mL ethanol, 50 mL of 1:10 diluted acetate buffer, and 50 μL 30% H_2O_2
- Acetate buffer: 27.2 g natriumacetate-trihydrate and 4.5 mL acetic acid added to 1 L distilled water (fine adjustment with acetic acid to a pH of 5.0)
- Electrode solutions: 1N H_3PO_4, 1N NaOH
- Electrofocusing strips
- Sample Application pieces
- Kerosene
- Nitrocellulose membrane (pore size 0.45 μm)
- Blotting paper
- Goat anti-human IgG
- Rabbit anti-goat IgG, HRP-labeled

All materials have to be stored as specified by the manufacturer (e.g., secondary antibodies at 4 °C). Blocking solution and antibody diluent should be prepared fresh on the day of IEF run as they contain milk.

1.2.2 Samples

CSF and serum are suitable for IEF. CSF is centrifuged at 2000 ×*g* for 10 min at room temperature before being used. Serum is isolated from blood by centrifugation after the blood samples were allowed to clot for ≥30 min [35].

1.2.3 Storage

IEF should be performed using fresh CSF and serum samples, but OCB may be even recovered after several years of storage at −20 °C [35] providing further evidence of the high stability of immunoglobulins [36].

1.3 Methods

1.3.1 Gel

IEF can be carried out using polyacrylamide [34] or agarose gels [37]. The gels have to be prepared using "carrier ampholytes" that build up a stable, linear pH gradient within the gel when subjected to an electric field. Typically, ampholytes with a pH range of 3–10 are used.

If using agarose, low electroendosmosis agarose is required because "ordinary" agarose (or even agar) is charged and therefore unsuitable for IEF [38]. Basically, large, high molecular weight proteins have a limited migration in polyacrylamide gels, but will move through agarose due to the greater pore size. Therefore, agarose type of gel is applied for proteins greater than 200 kDa [39]. For the detection of oligoclonal IgG bands, there is no relevant advantage of any of these two types of gel.

Following steps can be followed to produce a 7.5% polyacrylamide gel at a size of approximately 10 × 10 cm with a thickness of 2 mm. A total of 4.8 g urea is dissolved together with 5 mL of the 30% acrylamide solution and 1.5 mL carrier ampholyte in distilled water reaching a final volume of 20 mL (resulting in a final acrylamide concentration of 7.5%). After equilibrating the liquid gel in an ultrasonic bath for 15 min, 0.5 mL of the ammonium persulfate solution as well as 30 μL of tetramethylethylenediamine (TEMED) are added. The liquid gel is poured into the casting frame. Polymerization of the gel takes about 1 h. Until IEF run, the gel is set in a damp chamber at 4 °C.

1.3.2 Isoelectric Focusing

Preparation of Samples

CSF samples at IgG concentrations >3 mg/dL, as well as serum samples should be diluted in distilled water to achieve IgG concentrations of 3 mg/dL. If CSF samples show an IgG concentration < 3 mg/dL, CSF is applied without dilution to IEF and serum samples are diluted to the concentration of CSF (**Note 1**).

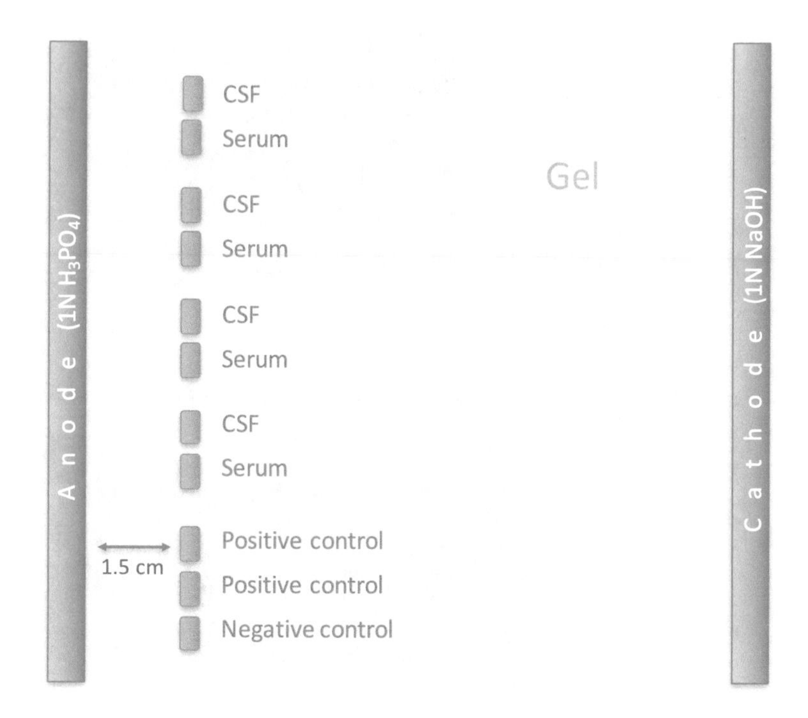

Fig. 3.4 Arrangement of filter papers on the gel for sample application at start of IEF. CSF, cerebrospinal fluid

Preparation of Electrophoresis Chamber and Sample Application

The gel is placed in the electrophoresis chamber, after 1 mL kerosene has been pipetted onto the cooling plate (**Note 2**). Then electrofocusing strips (filter paper) are placed onto the gel and soaked in electrolytes using 1N H_3PO_4 for the strip adjacent to the anode and 1N NaOH for the strip adjacent to the cathode. IEF sample application pieces (filter paper) are placed onto to the gel and 15 μL of (diluted) samples are added (Fig. 3.4, **Note 3**).

Run

IEF is carried out by applying a potential difference between the electrodes (1.08 kV, 15 mA, 200 W) for 2 h (**Note 4**).

1.3.3 Immunoblotting and -Staining

Mechanical Transfer

Gels are mechanically blotted on nitrocellulose membranes (**Note 5**). For this purpose, the gel is placed on a glass plate. On top of the gel, first nitrocellulose membrane is placed, followed by four blotting papers and another glass plate. This stack is weighted down by 1 kg for 20 min (Fig. 3.5).

Immunolabeling

After removing and discarding the blotters, nitrocellulose membrane is placed "protein side up" in blocking solution for 30 min at room temperature on a shaker. After that, the membrane is rinsed three times with 0.9% NaCl solution and incubated for 1 h with goat anti-human IgG at a final dilution of 1:2000 (in 50 mL antibody diluent paying attention that the gel is fully covered by the solution). Rinsing with tap water for ten times is followed by one

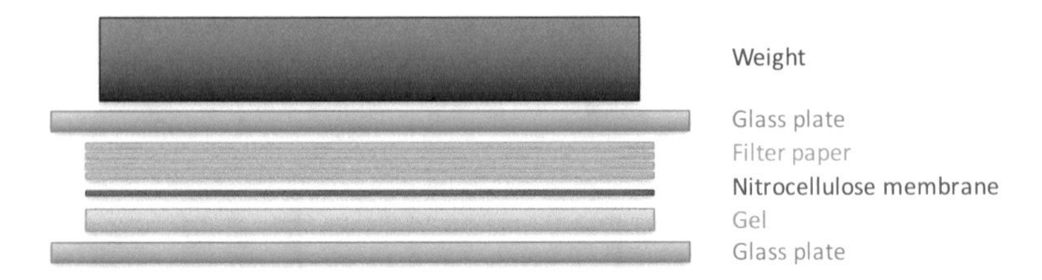

Fig. 3.5 Mechanical transfer of proteins to nitrocellulose membrane

Table 3.2
Recommendations for OCB detection in CSF and serum

1. Use IEF followed by immunodetection (e.g., blotting on nitrocellulose membrane and IgG-specific staining)
2. Use similar IgG concentration of applied CSF and serum samples[a]
3. Run CSF and serum sample in parallel, i.e., on the same gel
4. Include positive and negative controls on each gel

CSF cerebrospinal fluid, *IEF* isoelectric focusing, *OCB* oligoclonal bands
[a]CSF and serum samples are diluted as required, whereas concentrating CSF should be avoided as this leads to artifacts [8]

wash in antibody diluent for 5 min. Thereafter, membranes are incubated with horseradish peroxidase-labeled rabbit anti-goat IgG at a final dilution of 1:1000 in 50 mL antibody diluent for 1 h. Another rinsing with tap water for ten times is followed by one wash in 0.9% NaCl solution for 5 min.

Staining

Staining is performed by using 25 mg of 3-amino-9-ethylcarbazole diluted in 10 mL ethanol and 50 mL acetate buffer. After adding 50 μL of 30% hydrogen peroxide, membranes are incubated for 15 min. After development of the red-brown bands, membranes are washed with distilled water and air-dried (**Note 6**).

1.4 Notes

Detection of OCB is a technically demanding method. In the following, there are some notes that supplement the procedural steps explained in Subheading 1.3. Furthermore, Table 3.2 provides a list of red flags that should be considered in order to guarantee high validity of this method [8].

1.4.1 Note 1

IgG concentrations in CSF and serum are typically obtained by nephelometry in routine CSF diagnostics. Dilutions are made so that final IgG concentration of samples is 3 mg/dL with a variation of 0.1 mg/dL allowed before application to the gel.

Example 1: Serum IgG 985 mg/dL, CSF IgG 5.1 mg/dL.

To achieve a final concentration of 3 mg/dL, serum sample with an initial IgG concentration of 985 mg/dL would require a dilution factor of 1:328. For practicability reasons, we round dilution factors in steps of 10, i.e., in this example using 1:330, resulting in a final sample concentration of 2.98 mg/dL.

For CSF, a dilution of 1:1.7 has to be used to achieve a final concentration of 3 mg/dL.

Example 2: CSF IgG 2.1 mg/dL, Serum IgG 767 mg/dL

In this case, we use CSF without further dilution.

Serum is diluted in order to adjust IgG concentration to that of CSF. Division of serum IgG by CSF IgG results in a dilution factor of 365. After rounding up and using a dilution factor of 1:370, this leads to a final dilution of serum also to 2.1 mg/dL.

The application of 15–20 μL sample volume with a concentration of 3 mg/dL means that a total amount of 450–600 ng IgG is applied to the gel. This is the amount of IgG needed to achieve good, interpretable results [37].

1.4.2 Note 2

Before using the gel, it should equilibrate to room temperature. When placing the gel on the cooling plate of electrophoresis chamber, air bubbles have to be avoided in order to ensure sufficient contact to the cooling plate.

1.4.3 Note 3

Sample application pieces should be completely soaked with CSF and serum sample, respectively, otherwise the application of higher sample volume can be considered. Furthermore, there should be at least 1.5 cm distance between sample application pieces and the anodic strip (Fig. 3.4).

1.4.4 Note 4

Different protocols for IEF runs regarding voltage, current, power, and duration have been published depending on different considerations such as type of gel and the following aspects [37]:

During IEF, not only the proteins of the applied CSF and serum samples, but also the ampholytes are focused according to their isoelectric point, i.e., proteins and pH gradient migrate simultaneously. Voltage and focusing time determine the quality of OCB separation.

Using low volt-hours may lead to not adequately separated OCB, i.e., immunoglobulins have not migrated to their isoelectric point. Using "appropriate" volt-hours, the pH gradient is fully developed and the proteins have migrated properly. For reasons, which are still not clear, a phenomenon known as "cathodic drift" occurs with IEF in which the pH gradients slowly collapse into the cathode, when high volt-hours are applied. This means that some bands are lost into the catholyte (Fig. 3.6). This is the reason why IEF runs are fairly short.

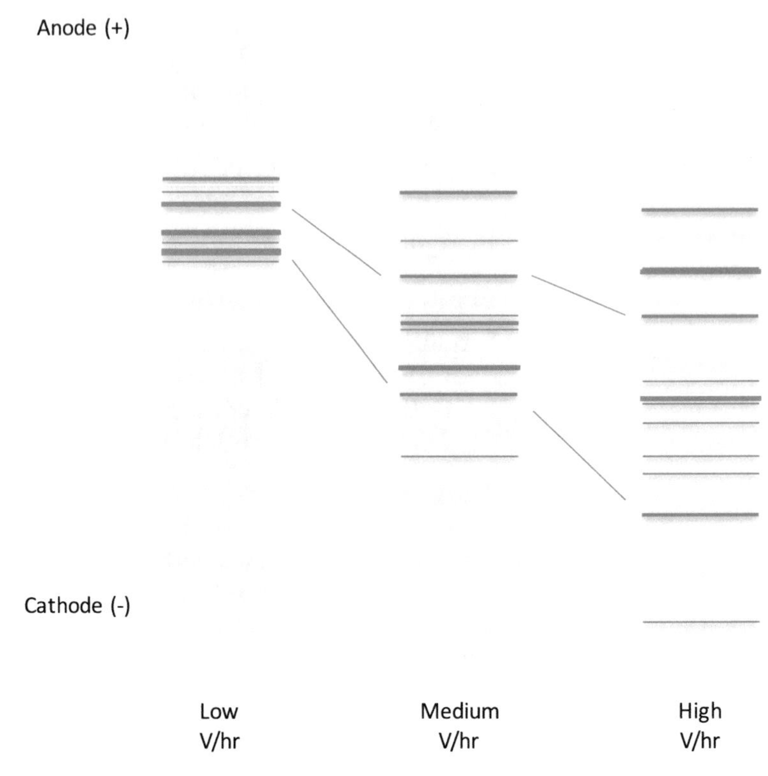

Fig. 3.6 Progress of isoelectric focusing dependent on duration of electrophoresis. Separation of oligoclonal IgG and focusing into sharp bands depending on volt-hours is shown. (Adapted after [37])

1.4.5 Note 5

After IEF, all filter papers are removed from the gel. Then, the anodic and cathodic end of the gel (where the filter strips were placed) are cut off, as these parts of the gel are usually deformed. Cutting these ends ensures that nitrocellulose membrane gets sufficient contact to the gel during blotting without riffles.

Place the nitrocellulose membrane carefully on the gel starting at one of the short ends and allow the membrane to fold gently onto the surface. Avoid trapping air underneath the membrane. At the end of blotting, nitrocellulose membrane has to be peeled off carefully. Furthermore, it is possible to do another blotting using another nitrocellulose membrane on the same gel.

1.4.6 Note 6

Once OCB are developed nitrocellulose membranes have to be protected from direct (and also indirect) sunlight, as they get bleached in rather short time.

2 κ Free Light Chains

2.1 Introduction

Besides intact immunoglobulins terminally differentiated B cells secrete an excess of FLC [40]. Both κ- and λ-FLC isotypes can be detected in the CSF and similar to other proteins such as the intact immunoglobulin molecule, FLC in CSF originate either from blood by diffusion across the blood–CSF barrier or are locally synthesized within the CNS [4]. For diagnostic purposes, determination of the locally synthesized fraction, i.e., distinct from the blood-derived fraction, is essential.

First, FLC are quantified in CSF and serum usually by nephelometry using antibodies against FLC-specific epitopes. For quantification of an intrathecal FLC synthesis, different approaches have been published among others the FLC index or more sophisticated nonlinear formulae [41].

In the following, we focus on κ-FLC, provide background on this molecule, describe the technology used for its determination, discuss how to interpret the results, and give insights in the potential application in clinical practice. The reason why we focus on κ-FLC is simply that most studies also focused on this isotype and presented favorable results of intrathecal κ-FLC synthesis in different neurological diseases, mostly multiple sclerosis, as compared to OCB testing.

2.1.1 κ Free Light Chains

Terminally differentiated B cells synthesize light chains and heavy chains that are finally bound together via disulfide bonds and non-covalent interactions in the endoplasmic reticulum to form a complete immunoglobulin (see Subheading 1.1.1). Light chains have a molecular weight of approximately 24 kD and consists of two immunoglobulin domains, a constant region (C_L) that specifies the isotype of free light chain (either κ or λ) and a variable domain (V_L) that is part of the antigen-binding site (Fig. 3.1) [9]. However, light chains are produced in 10–40% excess over heavy chains and are secreted as free form into blood circulation. FLC exist in the form of monomers, non-covalent, or covalent dimers. κ-FLC consist mainly of monomers but also of non-covalent dimers; λ-FLC are present as covalent dimers. The functions of FLC is not fully elucidated, but it seems that FLC have only little or no antigen-binding activity [40].

2.1.2 History of Free Light Chain Detection

After Henry Bence Jones in 1847 described a protein in the urine of a patient with severe bone pain and fractures that precipitated by addition of nitric acid [42], determination of Bence Jones protein evolved to an important diagnostic marker for patients with multiple myeloma. More than 100 years after its discovery, the Bence Jones protein was identified as monoclonal FLC [43]. Further progress in laboratory methods to detect FLC not only in urine, but also in serum included protein electrophoresis and immunofixation electrophoresis. However, these methods showed only limited sensitivity (serum protein electrophoresis: 500–2.000 mg/L, immunofixation electrophoresis: 150–500 mg/L) so that low level FLC, e.g., under physiological or oligosecretory conditions were not detectable and, furthermore, allowed only qualitative determination [43].

Attempts to quantify FLC were initially hindered by difficulties of producing antibodies specific to FLCs that do not cross-react with light chains bound in intact immunoglobulins. The breakthrough was achieved in 2001 by Bradwell and coworkers who dissociated light chains from heavy chains and then raised antibodies directed against unique epitopes on FLC that are normally "hidden" in the conformational structure of an intact immunoglobulin [44]. These anti-human FLC-specific antibodies could then be used to develop assays that can exclusively detect FLC at least a hundred times more sensitive than previous methods with detection limits down to approximately 1 mg/L. Recently, FLC assay have been developed using monoclonal instead of polyclonal detection antibodies with similar diagnostic performance [45].

2.1.3 Clinical Application

Determination of FLC is the mainstay in diagnosis, prognosis, and monitoring of patients with monoclonal plasma-proliferative diseases—reaching from monoclonal gammopathy of undetermined significance (MGUS) to multiple myeloma and light chain amyloidosis—and currently recommended by the guidelines of the International Myeloma Working group [46, 47].

In recent years, polyclonal FLC have gained increasing attention from the scientific community and were determined in various autoimmune diseases. Increased serum levels of FLC have been observed in systemic lupus erythematosus, rheumatoid arthritis, and primary Sjögren's syndrome [48, 49]. In the field of neurology, the majority of studies focused on patients with multiple sclerosis and showed diagnostic accuracy of intrathecal κ-FLC similar to that of OCB.

Various approaches have been suggested to determine an intrathecal FLC synthesis:

- CSF κ-FLC/CSF total protein quotient [50]
- $Q_{\kappa\text{-FLC}}$ (CSF κ-FLC/serum κ-FLC quotient) [51–53]
- κ-FLC index ([CSF κ-FLC/ serum κ-FLC]/[CSF albumin/ serum albumin]) [54–56]
- $IF_{\kappa\text{-FLC}}$—intrathecal κ-FLC fraction. This method refers $Q_{\kappa\text{-FLC}}$ to Q_{alb} (CSF albumin/ serum albumin) in a control population and calculates a Q_{alb}-dependent upper normal ($Q_{lim\ \kappa\text{-FLC}}$), i.e., considers normal diffusion of κ-FLC through the blood–CSF barrier. Relating a patient's individual $Q_{\kappa\text{-FLC}}$ to $Q_{lim\ \kappa\text{-FLC}}$

reveals whether an intrathecal synthesis is present or not and also allows to express a percentage intrathecal fraction (local κ-FLC synthesis as percentage of total CSF κ-FLC concentration) [56].

There are no conclusive data which approach reveals the highest diagnostic accuracy. However, it seems that $IF_{\kappa\text{-FLC}}$ is similar to κ-FLC index [56], $Q_{\kappa\text{-FLC}}$ similar to CSF κ-FLC levels [51] and that κ-FLC index is superior to CSF κ-FLC [54]. Although CSF κ-FLC shows already a high diagnostic accuracy due to the relatively strong elevated levels in case of intrathecal inflammation, application of any formulae that corrects for the blood-derived fraction probably adds some additional value. Further studies comparing these different approaches within one patient cohort are still needed.

2.2 Materials

2.2.1 Assays

κ-FLC can be measured by different laboratory methods including immunoturbidimetry/nephelometry or enzyme-linked immunosorbent assay (ELISA) using commercially available test kits containing anti-human κ-FLC-specific antibodies. In the following, we focus on nephelometry for κ-FLC detection, describe the characteristics and limitations of this method especially with regard to the different types of detection antibodies available (Subheading 2.3).

2.2.2 Samples

CSF and serum can be used for immunoturbidimetric and nephelometric assays. Instead of serum also lithium-heparin plasma and citrated plasma can be used [57, 58].

2.2.3 Storage

Specimens are stable for at least 1 week when stored refrigerated at 2–8 °C or longer when frozen at −20 °C and −80 °C [57].

2.3 Methods

2.3.1 Principles

Nephelometry

Evaluation of the concentration of κ-FLC (antigen) by nephelometry requires the addition of a solution with detection antibodies to a cuvette containing CSF or serum sample. Then, a beam of monochrome light is passed through the cuvette. As the antigen–antibody reaction proceeds, light is scattered as insoluble immune complexes are formed. Light scatter is determined by measuring light intensity at a 90° angle away from the incident light (Fig. 3.7). For the formation of light-scattering immune complexes, an optimal ratio between antigen and antibody molecules is required (Fig. 3.8). This is when the detection antibody in the cuvette is in moderate excess. Then, the number of immune complexes formed is proportional to the antigen (κ-FLC) concentration. In this context, testing for antigen excess is required in order to avoid false low results (see Subheading 2.4.4).

A series of calibrators of known antigen concentration are assayed initially to produce a calibration curve of measured light scatter versus antigen concentration. Patients' samples of unknown antigen concentration can then be measured and the results read

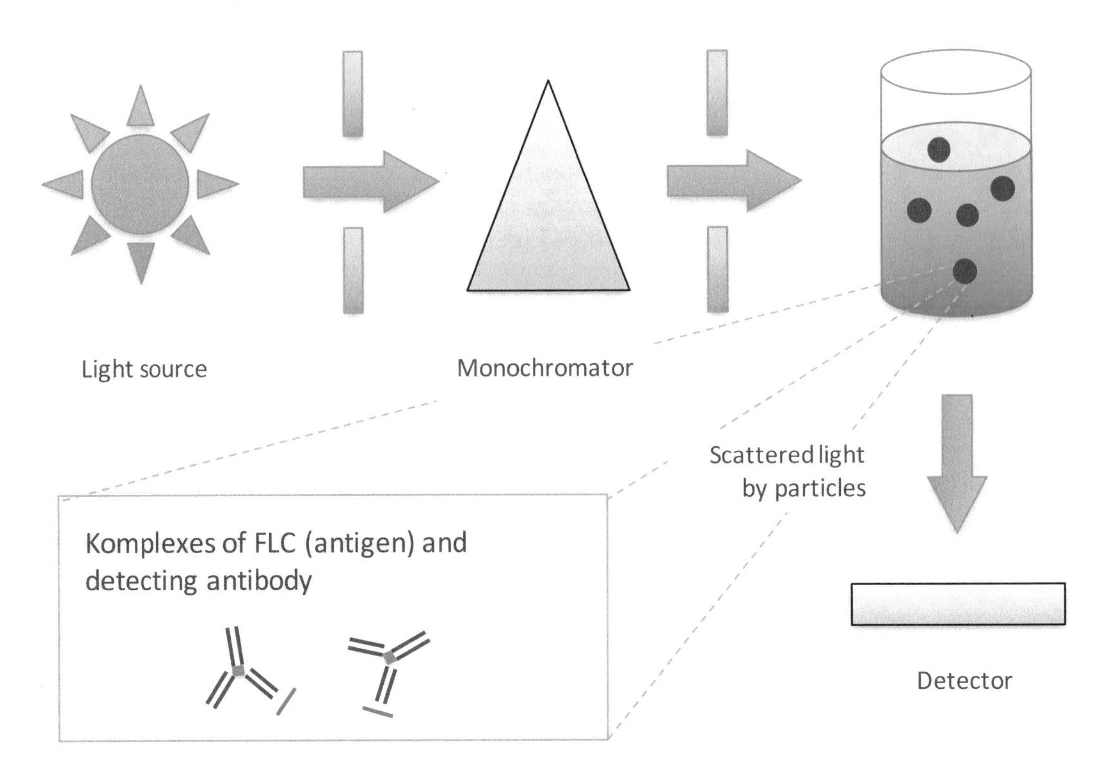

Fig. 3.7 Principles of FLC detection by nephelometry. A beam of light is passed through a cuvette and increasingly scattered as immune complexes consisting of κ-FLC and detection antibodies are formed. Referring intensity of scattered light to a calibration curve reveals the concentrations of κ-FLC. *FLC* free light chain

from the calibration curve. Generally, two types of kinetic measurements can be used, either fixed-time or peak-rate nephelometry. Details on basic principles of nephelometric measurement can be found elsewhere [59].

κ-FLC Assays

There are two main types of assays that use either polyclonal or monoclonal detection antibodies directed against human κ-FLC.

Polyclonal detection antibodies are raised in sheep after immunization with a pool of human κ-FLC purified from urine of patients containing Bence Jones protein. The resulting antisera are adsorbed against intact immunoglobulin proteins (containing bound light chains), so that only κ-FLC-specific antibodies remain in the final antisera [44].

Monoclonal anti-human FLC antibodies are produced in mice. Mice are immunized with Bence Jones protein purified from patients' urine. Then, their spleens are removed to isolate B cells and to fuse them with myeloma cells. Out of these hybrid cells, those producing antibodies against κ-FLC are selected. For the final reagent, different batches of monoclonal antibodies are mixed together [45].

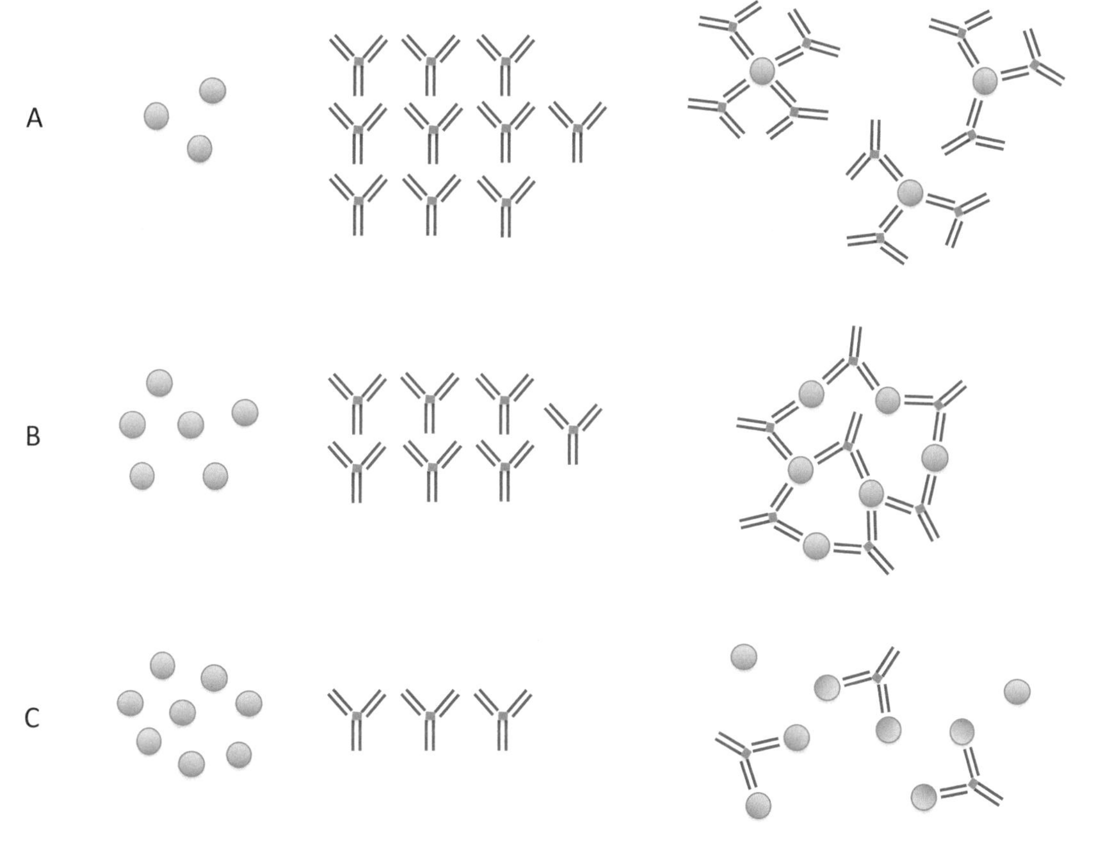

Fig. 3.8 Antigen–antibody reaction at different antigen and antibody concentrations. (**a**) Antibody excess. (**b**) Optimal ratio between antibodies and antigen. (**c**) Antigen excess

The main characteristic of both the poly- and monoclonal detection antibodies is that they recognize light chain epitopes that are hidden when light chains are bound to the intact immunoglobulin molecule, but are exposed when light chains circulate freely. Whereas polyclonal antibodies are able to recognize a multitude of variations within the light chain protein per se, monoclonal antibodies are highly specific to one target. That is why, in the latter case, a panel of different monoclonal antibodies is used in the final assay reagent. Furthermore, detection antibodies are conjugated with latex beads to increase sensitivity of nephelometric assay, as linking the antibody to a suitably sized particle increases the relative light-scattering signal of the antigen–antibody reaction [57, 58]. Differences between monoclonal and polyclonal detection antibodies are discussed below and summarized in Table 3.3.

2.4 Notes

There are some characteristics of nephelometric κ-FLC measurement that apply in general and some that specifically differ between the two types of assay (using polyclonal vs. monoclonal anti-human κ-FLC antibodies). In the following, the main issues are discussed

Table 3.3
Characteristics of and differences between nephelometric FLC assay

	Assay using anti-human FLC polyclonal detection antibody (1)	Assay using anti-human FLC monoclonal detection antibody (2)
Detection antibody	Polyclonal anti-human κ-FLC latex-conjugated antibodies prepared by immunization of, e.g., sheep with a pool of human FLC	A panel of monoclonal anti-human κ-FLC latex-conjugated antibodies raised in mice
Antigen addressed	Epitopes on poly- and monoclonal human κ-light chain that are hidden if light chain is bound within the immunoglobulin molecule, but accessible if light chain is free	
Cross-reactivity [a]	<0.05%	<0.01%
Detection limit	1 mg/L	
Lot-to-lot variation	Higher ~15–20%	Lower ~7%
Interference	– Highly lipemic and grossly hemolyzed samples might lead to falsely high results – Extreme polymerization of FLC can cause overestimation (by as much as ten-fold)—occurs very rarely	
Nonlinearity	In presence of high concentrations of monoclonal intact immunoglobulins	
	In case of high polyclonal κ-FLC concentrations	
Non-reactivity	Changes in amino acid sequence of the light chain may render certain light chain epitopes unrecognizable to the κ-FLC reagents—occurs very rarely	
Reference values	Exist only for serum (and were determined in healthy controls as 95% reference interval)	
	3.3–19.4 mg/L[b]	6.7–22.4 mg/L[c]
	For determination of an intrathecal κ-FLC synthesis, different approaches, e.g., calculation of the κ-FLC index, have been applied and for κ-FLC index similar cut-offs were published (approx. 7) [d]	

Specifications and data displayed in the left column (1) apply to the Freelite assay (The Binding Site, Birmingham, UK), and in the right column (2) to the N Latex FLC assay (Siemens, Erlangen, Germany)

[a]Cross-reactivity of detection antibodies between free and bound κ-light chain (percentages are given in case of adding purified IgG)

[b][60]

[c][45]

[d][34, 54, 61]

and summarized in Table 3.3. For detailed information, refer to the manuals of the manufacturer, as these are commercially available assays.

2.4.1 Cross-Reactivity of Anti-κ-FLC Antibodies

As intact immunoglobulins are present in CSF and serum, bound κ-light chain can potentially cross-react in κ-FLC assay and lead to an overestimation of κ-FLC concentrations, i.e., to falsely high values. Although specificity of detection antibodies is directed to residues on the C_L domain of κ-light chains that are only exposed if circulating freely, there is a cross-reactivity of <0.05% at normal serum immunoglobulin levels [57, 62].

2.4.2 Lot-to-Lot Variation

Amino acid sequence variation within the primary structure of κ-light chain leads to a heterogeneous mixture of κ-light chain molecules and, therefore, also in variation of κ-FLC-specific epitopes. Due to this variability, different batches of anti-κ-FLC antibodies may react differently to the addressed epitopes in individual patients. In these instances, sample results may vary when tested using multiple batches. The lot-to-lot variation is typically higher in assays using polyclonal antibodies (approximately 15%), as new batches are the result of renewed immunization of, e.g., sheep (potentially with another pool of κ-FLC) and purification [63]. In monoclonal antibodies that are always raised against the same preselected and unique epitope, lot-to-lot variation is smaller (around 7%) [64]. With respect to clinical routine, reduced batch variation is important for patients monitoring over time, i.e., when samples are collected longitudinally and therefore, are not measured using the same batch of detection antibodies.

2.4.3 Interference

Nephelometric assays are not suitable for measurement of highly lipemic or grossly hemolyzed samples or samples containing high levels of circulating immune complexes due to the unpredictable degree of nonspecific scatter these sample types may generate. Which hemoglobin and triglyceride concentrations are still acceptable is provided by the manufacturer for each FLC assay and platform. For example, using the Freelite assay (The Binding Site), an interference of approximately −6% by 200 mg/L bilirubin and +8% by 5 g/L hemoglobin has been demonstrated in control serum at a concentration of 40 mg/L κ-FLC. In case of unexpected results, it should be confirmed using an alternative method.

Furthermore, polymeric forms can also cause an overestimation of FLC possibly by reaction at multiple antigenic sites on the FLC molecule. The presence of multimeric forms in addition to the usual κ-FLC monomers has been reported in sera from patients with nonsecretory multiple myeloma in a few cases [46, 57].

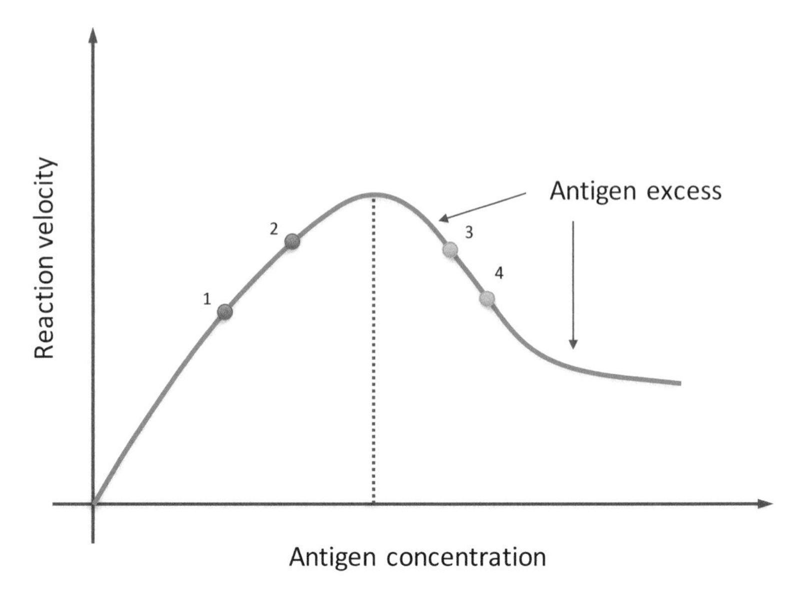

Fig. 3.9 Detection of antigen excess. *Scenario A (red):* After first measurement (1) of the sample containing antigen and detection antibodies, calibrator (i.e., further antigen) is added to the cuvette for second measurement (2). Further increase in reaction velocity indicates that there is no antigen excess. *Scenario B (orange):* The lack of an increase in reaction velocity in the second measurement (4) as compared to the first measurement (3) indicates antigen excess. Therefore, the original sample has to be further diluted and another measurement performed (including another testing for antigen excess)

2.4.4 Antigen Excess

Antigen excess is uncommon, however, a small proportion of samples containing very high concentrations of κ-FLC can give falsely low results due to antigen excess.

Antigen excess is independent of the assay used (i.e., type of detection antibody). One approach not missing antigen excess is the use of pre-reaction protocols (as applied for fixed-time nephelometry). First, a small amount of sample is added to the reaction cuvette that should give sufficient agglutination, before the rest of the sample is applied. If the signal at the end of pre-reaction is above a certain threshold, another analysis of the sample using the next higher dilution is started [45, 58]. Another possibility is to do a second measurement under certain circumstances (as applied for peak-rate nephelometry), e.g., when the first measurement—using the standard required amount of sample—is beyond a certain value. For second measurement, calibrator (i.e., further antigen) is added to the cuvette. A lack of increase in reaction velocity in the second measurement indicates antigen excess (Fig. 3.9). In that case, the original sample is diluted and another measurement is performed (including another testing for antigen excess) [59].

2.4.5 Nonlinearity

In general, FLC dilute in a linear manner. However, certain monoclonal FLC samples show a nonlinear dilution response, i.e., the result in the next dilution provide a higher value than expected.

κ-FLC are hindered in the presence of high concentrations of monoclonal intact immunoglobulins. This effect is patient-dependent and independent of the assay used. In patients with high polyclonal κ-FLC concentrations, e.g., due to renal failure, possible false results at the initial dilution can be obtained using the polyclonal assay [65].

Nonlinear samples require several sample dilutions to reach a final κ-FLC concentration. Therefore, to prevent effect of nonlinearity in the polyclonal assay, it is recommended to use either serial dilutions or to re-run samples at the next higher dilution as long as the κ-FLC concentrations are above a certain level (e.g., κ-FLC $>$ 50 mg/L). For the monoclonal assay, second analyses at a higher dilution are recommended for samples containing high concentrations of intact immunoglobulins [57, 58].

2.4.6 Non-reactivity

Each monoclonal κ-FLC contains unique amino acid combinations. It is theoretically possible for certain κ-FLC molecules to be undetectable by immunoassay leading to lower than expected measurements. In practice, this occurs rarely and has been reported in patients with amyloidosis and after changing of reagent lot in multiple myeloma [63, 66]. Suspected samples (with suspected falsely low κ-FLC concentrations) should first be tested for antigen excess (*see* Subheading 2.4.4) and then further investigated by other laboratory methods (e.g., immunofixation, serum protein electrophoresis).

References

1. Reiber H, Peter JB (2001) Cerebrospinal fluid analysis: disease-related data patterns and evaluation programs. J Neurol Sci 184 (2):101–122
2. Deisenhammer F, Bartos A, Egg R, Gilhus NE, Giovannoni G, Rauer S et al (2006) Guidelines on routine cerebrospinal fluid analysis. Report from an EFNS task force. Eur J Neurol 13 (9):913–922
3. Thompson AJ, Banwell BL, Barkhof F, Carroll WM, Coetzee T, Comi G et al (2018) Diagnosis of multiple sclerosis: 2017 revisions of the McDonald criteria. Lancet Neurol 17 (2):162–173
4. Reiber H (2003) Proteins in cerebrospinal fluid and blood: barriers, CSF flow rate and source-related dynamics. Restor Neurol Neurosci 21 (3–4):79–96
5. Link H, Tibbling G (1977) Principles of albumin and IgG analyses in neurological disorders. III. Evaluation of IgG synthesis within the central nervous system in multiple sclerosis. Scand J Clin Lab Invest 37(5):397–401
6. Reiber H (1994) Flow rate of cerebrospinal fluid (CSF)—a concept common to normal blood-CSF barrier function and to dysfunction in neurological diseases. J Neurol Sci 122 (2):189–203
7. Auer M, Hegen H, Zeileis A, Deisenhammer F (2016) Quantitation of intrathecal immunoglobulin synthesis—a new empirical formula. Eur J Neurol 23(4):713–721
8. Freedman MS, Thompson EJ, Deisenhammer F, Giovannoni G, Grimsley G, Keir G et al (2005) Recommended standard of cerebrospinal fluid analysis in the diagnosis of

multiple sclerosis: a consensus statement. Arch Neurol 62:865–870
9. Abbas AK, Lichtman AH, Pillai S Cellular and molecular immunology, 6th edn. Saunders Elsevier, Philadelphia
10. Lange C (1912) Die Ausflockung kolloidalen Goldes durch Zerebrospinalflüssigheit bei luetischen Affektionen des Zentralnervensystems. Z Chemother 1:44–78
11. Kabat EM, Moore DH, Landow H (1942) An electrophoretic study of the protein components in cerebrospinal fluid and their relationship to the serum proteins. J Clin Invest 21 (5):571–577
12. Katzenelbogen S (1935) The cerebrospinal fluid and its relation to the blood. John Hopkins, Baltimore
13. Frick E (1959) [Immunophoretic studies on cerebrospinal fluid]. Klin Wochenschr. 37 (12):645–651
14. Link H (1967) Immunoglobulin G and low molecular weight proteins in human cerebrospinal fluid. Chemical and immunological characterisation with special reference to multiple sclerosis. Acta Neurol Scand 43(Suppl 28):1–136
15. Delmotte P (1972) [Comparative results of agar electrophoresis and electrofocalization examination of gamma globulins of the cerebrospinal fluid]. Acta Neurol Belg. 72 (4):226–234
16. Walker RW, Keir G, Johnson MH, Thompson EJ (1983) A rapid method for detecting oligoclonal IgG in unconcentrated CSF, by agarose isoelectric focusing, transfer to cellulose nitrate and immunoperoxidase staining. J Neuroimmunol 4(2):141–148
17. Brasher MD, Thorpe R (1998) Isoelectric focusing. In: Delves PJ, Roitt IM, eds Encyclopedia of immunology. 2nd ed.
18. Andersson M, Alvarez-Cermeño J, Bernardi G, Cogato I, Fredman P, Frederiksen J et al (1994) Cerebrospinal fluid in the diagnosis of multiple sclerosis: a consensus report. J Neurol Neurosurg Psychiatry 57(8):897–902
19. Kostulas VK, Link H, Lefvert AK (1987) Oligoclonal IgG bands in cerebrospinal fluid. Principles for demonstration and interpretation based on findings in 1114 neurological patients. Arch Neurol 44(10):1041–1044
20. Ohman S, Ernerudh J, Forsberg P, Henriksson A, von Schenck H, Vrethem M (1992) Comparison of seven formulae and isoelectrofocusing for determination of intrathecally produced IgG in neurological diseases. Ann Clin Biochem 29(Pt 4):405–410
21. McLean BN, Luxton RW, Thompson EJ (1990) A study of immunoglobulin G in the cerebrospinal fluid of 1007 patients with suspected neurological disease using isoelectric focusing and the Log IgG-index. A comparison and diagnostic applications. Brain 113 (Pt 5):1269–1289
22. Haghighi S, Andersen O, Rosengren L, Bergström T, Wahlström J, Nilsson S (2000) Incidence of CSF abnormalities in siblings of multiple sclerosis patients and unrelated controls. J Neurol 247(8):616–622
23. Hegen H, Auer M, Zeileis A, Deisenhammer F (2016) Upper reference limits for cerebrospinal fluid total protein and albumin quotient based on a large cohort of control patients: implications for increased clinical specificity. Clin Chem Lab Med 54(2):285–292
24. Davies G, Keir G, Thompson EJ, Giovannoni G (2003) The clinical significance of an intrathecal monoclonal immunoglobulin band: a follow-up study. Neurology 60(7):1163–1166
25. Bourahoui A, de Sèze J, Guttierez R, Onraed B, Hennache B, Ferriby D et al (2004) CSF isoelectrofocusing in a large cohort of MS and other neurological diseases. Eur J Neurol 11(8):525–529
26. Compston A, Coles A (2002) Multiple sclerosis. Lancet 359(9313):1221–1231
27. Aboul-Enein F, Seifert-Held T, Mader S, Kuenz B, Lutterotti A, Rauschka H et al (2013) Neuromyelitis optica in Austria in 2011: to bridge the gap between neuroepidemiological research and practice in a study population of 8.4 million people. PLoS One 8(11): e79649
28. Jarius S, Paul F, Franciotta D, Ruprecht K, Ringelstein M, Bergamaschi R et al (2011) Cerebrospinal fluid findings in aquaporin-4 antibody positive neuromyelitis optica: results from 211 lumbar punctures. J Neurol Sci 306 (1–2):82–90
29. Menge T, Hemmer B, Nessler S, Wiendl H, Neuhaus O, Hartung H-P et al (2005) Acute disseminated encephalomyelitis: an update. Arch Neurol Am Med Assoc 62 (11):1673–1680
30. Burgoon MP, Hammack BN, Owens GP, Maybach AL, Eikelenboom MJ, Gilden DH (2003) Oligoclonal immunoglobulins in cerebrospinal fluid during varicella zoster virus (VZV) vasculopathy are directed against VZV. Ann Neurol 54(4):459–463
31. Link H, Huang Y-M (2006) Oligoclonal bands in multiple sclerosis cerebrospinal fluid: an update on methodology and clinical usefulness. J Neuroimmunol 180(1–2):17–28

32. Sellebjerg F, Christiansen M (1996) Qualitative assessment of intrathecal IgG synthesis by isoelectric focusing and immunodetection: interlaboratory reproducibility and interobserver agreement. Scand J Clin Lab Invest 56 (2):135–143
33. Sellebjerg F, Christiansen M, Rasmussen LS, Jaliachvili I, Nielsen PM, Frederiksen JL (1996) The cerebrospinal fluid in multiple sclerosis. Quantitative assessment of intrathecal immunoglobulin synthesis by empirical formulae. Eur J Neurol 3:548–559
34. Hegen H, Milosavljevic D, Schnabl C, Manowiecka A, Walde J, Deisenhammer F et al (2018) Cerebrospinal fluid free light chains as diagnostic biomarker in neuroborreliosis. Clin Chem Lab Med 56:1383
35. Teunissen CE, Petzold A, Bennett JL, Berven FS, Brundin L, Comabella M et al (2009) A consensus protocol for the standardization of cerebrospinal fluid collection and biobanking. Neurology 73(22):1914–1922
36. Gislefoss RE, Grimsrud TK, Mørkrid L (2009) Stability of selected serum proteins after long-term storage in the Janus Serum Bank. Clin Chem Lab Med 47(5):596–603
37. Keir G, Luxton RW, Thompson EJ (1990) Isoelectric focusing of cerebrospinal fluid immunoglobulin G: an annotated update. Ann Clin Biochem 27(Pt 5):436–443
38. Guo Y, Li X, Fang Y (1998) The effects of electroendosmosis in agarose electrophoresis. Electrophoresis 19(8–9):1311–1313
39. Herbert B (2005) Some practices and pitfalls of sample preparation for isoelectric focusing in proteomics. In: Garfin D, Ahuja S, eds. Handbook of isoelectric focusing and proteomics. pp 147–164
40. Nakano T, Matsui M, Inoue I, Awata T, Katayama S, Murakoshi T (2011) Free immunoglobulin light chain: its biology and implications in diseases. Clin Chim Acta 412 (11–12):843–849
41. Ramsden DB (2017) Multiple sclerosis: assay of free immunoglobulin light chains. Ann Clin Biochem 54(1):5–13
42. Jones HB (1847) Chemical pathology. Lancet 2:88–92
43. Jenner E (2014) Serum free light chains in clinical laboratory diagnostics. Clin Chim Acta 427:15–20
44. Bradwell AR, Carr-Smith HD, Mead GP, Tang LX, Showell PJ, Drayson MT et al (2001) Highly sensitive, automated immunoassay for immunoglobulin free light chains in serum and urine. Clin Chem 47(4):673–680
45. Te Velthuis H, Knop I, Stam P, van den Broek M, Bos HK, Hol S et al (2011) N latex FLC—new monoclonal high-performance assays for the determination of free light chain kappa and lambda. Clin Chem Lab Med 49 (8):1323–1332
46. Dispenzieri A, Kyle R, Merlini G, Miguel JS, Ludwig H, Hajek R et al (2009) International Myeloma Working Group guidelines for serum-free light chain analysis in multiple myeloma and related disorders. Leukemia 23:215–224
47. Rajkumar SV, Dimopoulos MA, Palumbo A, Blade J, Merlini G, Mateos M-V et al (2014) International myeloma working group updated criteria for the diagnosis of multiple myeloma. Lancet Oncol 15(12):e538–e548
48. Aggarwal R, Sequeira W, Kokebie R, Mikolaitis RA, Fogg L, Finnegan A et al (2011) Serum free light chains as biomarkers for systemic lupus erythematosus disease activity. Arthritis Care Res 63(6):891–898
49. Gottenberg J-E, Aucouturier F, Goetz J, Sordet C, Jahn I, Busson M et al (2007) Serum immunoglobulin free light chain assessment in rheumatoid arthritis and primary Sjogren's syndrome. Ann Rheum Dis 66(1):23–27
50. Hassan-Smith G, Durant L, Tsentemeidou A, Assi LK, Faint JM, Kalra S et al (2014) High sensitivity and specificity of elevated cerebrospinal fluid kappa free light chains in suspected multiple sclerosis. J Neuroimmunol 276 (1–2):175–179
51. Makshakov G, Nazarov V, Kochetova O, Surkova E, Lapin S, Evdoshenko E (2015) Diagnostic and prognostic value of the cerebrospinal fluid concentration of immunoglobulin free light chains in clinically isolated syndrome with conversion to multiple sclerosis. PLoS One 10(11):e0143375
52. Senel M, Tumani H, Lauda F, Presslauer S, Mojib-Yezdani R, Otto M et al (2014) Cerebrospinal fluid immunoglobulin kappa light chain in clinically isolated syndrome and multiple sclerosis. PLoS One 9(4):e88680
53. Fischer C, Arneth B, Koehler J, Lotz J, Lackner KJ (2004) Kappa free light chains in cerebrospinal fluid as markers of intrathecal immunoglobulin synthesis. Clin Chem 50 (10):1809–1813
54. Presslauer S, Milosavljevic D, Brücke T, Bayer P, Hübl W, Hübl W (2008) Elevated levels of kappa free light chains in CSF support the diagnosis of multiple sclerosis. J Neurol 255(10):1508–1514
55. Presslauer S, Milosavljevic D, Huebl W, Parigger S, Schneider-Koch G, Bruecke T

(2014) Kappa free light chains: diagnostic and prognostic relevance in MS and CIS. PLoS One 9(2):e89945

56. Presslauer S, Milosavljevic D, Huebl W, Aboulenein-Djamshidian F, Krugluger W, Deisenhammer F et al (2016) Validation of kappa free light chains as a diagnostic biomarker in multiple sclerosis and clinically isolated syndrome: a multicenter study. Mult Scler 22 (4):502–510
57. Tate J, Bazeley S, Sykes S, Mollee P (2009) Quantitative serum free light chain assay—analytical issues. Clin Biochem Rev 30 (3):131–140
58. Te Velthuis H, Drayson M, Campbell JP (2016) Measurement of free light chains with assays based on monoclonal antibodies. Clin Chem Lab Med 54(6):1005–1014
59. Thomas L (1990) Quantitative determination of plasma proteins by antigen-antibody reactions using nephelometry and turbidimetry. J Lab Med 14:313–320
60. Katzmann JA, Clark RJ, Abraham RS, Bryant S, Lymp JF, Bradwell AR et al (2002) Serum reference intervals and diagnostic ranges for free kappa and free lambda immunoglobulin light chains: relative sensitivity for detection of monoclonal light chains. Clin Chem 48 (9):1437–1444
61. Leurs C, Twaalfhoven H, Witte BI, van Pesch V, Dujmovic I, Drulovic J et al (2017) Kappa free light chains: an automated alternative to oligoclonal bands for CIS and MS diagnosis? Multiple Scler J 23(S3):591–593
62. Robson E, Mead G, Bradwell A (2006) To the editor: in reply to Nakano et al. Clin Chem Lab Med 44(5):522–532. Clin Chem Lab Med. 2007;45(2):264–5; authorreply 266–7
63. Tate JR, Mollee P, Dimeski G, Carter AC, Gill D (2007) Analytical performance of serum free light-chain assay during monitoring of patients with monoclonal light-chain diseases. Clin Chim Acta 376(1–2):30–36
64. Pretorius CJ, Klingberg S, Tate J, Wilgen U, Ungerer JPJ (2012) Evaluation of the N latex FLC free light chain assay on the Siemens BN analyser: precision, agreement, linearity and variation between reagent lots. Ann Clin Biochem 49(Pt 5):450–455
65. Jacobs JFM, Hoedemakers RMJ, Teunissen E, van der Molen RG, Te Velthuis H (2012) Effect of sample dilution on two free light chain nephelometric assays. Clin Chim Acta 413(19–20):1708–1709
66. Palladini G, Russo P, Bosoni T, Verga L, Sarais G, Lavatelli F et al (2009) Identification of amyloidogenic light chains requires the combination of serum-free light chain assay with immunofixation of serum and urine. Clin Chem 55(3):499–504

Chapter 4

Abeta CSF LC-MS

Josef Pannee, Kaj Blennow, and Henrik Zetterberg

Abstract

In this chapter, a method using an antibody independent approach based on solid phase extraction (SPE) and liquid chromatography (LC)-tandem mass spectrometry (MS/MS) is described for the analysis of Aβ isoforms in cerebrospinal fluid (CSF). Stable isotope-labeled Aβ peptides are used as internal standards, enabling absolute quantification. A high-resolution quadrupole-Orbitrap hybrid instrument was used for measurements. The method allows quantification of CSF Aβ1-42 between 150 and 4000 pg/mL.

Key words Alzheimer's disease, Amyloid beta peptides, Cerebrospinal fluid, Solid phase extraction, Mass spectrometry, Liquid chromatography

1 Introduction

Alzheimer's disease (AD) is a neurodegenerative disease which together with other dementias is estimated to effect over 46 million people globally. AD, which accounts for 50–70% of dementia cases [1] is characterized by neurodegeneration—synaptic degeneration and loss, plaques consisting of aggregated amyloid β (Aβ) and neurofibrillary tangles composed of aggregated and phosphorylated tau protein [2]. Patients with AD have a decreased concentration of the 42-amino acid form of Aβ ($A\beta_{1-42}$) in cerebrospinal fluid (CSF), which likely reflects cortical Aβ deposition [3]. This biomarker, together with increased CSF total tau (T-tau) and phosphorylated tau (P-tau) concentrations, is now an established core biomarker for AD [4]. In line with this, assessment of CSF biomarkers and plaque pathology in vivo have recently been included in the research diagnostic criteria for AD [5]. For CSF measurements of $A\beta_{1-42}$, several immunoassays are available and used in many clinical laboratories [4].

These immunoassays, i.e., antibody based techniques, may be influenced by matrix effects. The use of immunoassays on different technology platforms and lack of assay standardization [6–8] has made the introduction of global cut-off concentrations difficult

Charlotte E. Teunissen and Henrik Zetterberg (eds.), *Cerebrospinal Fluid Biomarkers*, Neuromethods, vol. 168, https://doi.org/10.1007/978-1-0716-1319-1_4,

[9, 10]. An analytically validated reference measurement procedure (RMP) would enable uniform calibration of different assay platforms and ideally result in better comparability across analytical platforms.

We here describe absolute quantification of $A\beta_{1\text{-}42}$ using liquid chromatography-tandem mass spectrometry (LC-MS/MS). The method described is listed as an RMP by the Joint Committee for Traceability in Laboratory medicine (JCTLM database identification number C11RMP9) [11] and is used to determine the absolute concentration of $A\beta_{1\text{-}42}$ in a Certified Reference Material (CRM) to harmonize CSF $A\beta_{1\text{-}42}$ measurements across techniques and analytical platforms. The workflow described here could be of relevance for the development of candidate reference methods for peptides and proteins within other areas of medicine.

For the method described in this chapter, $A\beta_{1\text{-}42}$ is extracted from CSF using solid phase extraction. The extracted samples are then injected on an LC and separated over a reversed-phase monolithic column using basic mobile phases. Following LC separation, targeted quantification is performed using a quadrupole-Orbitrap hybrid mass spectrometer operated in the parallel reaction monitoring (PRM) mode. In PRM, $A\beta_{1\text{-}42}$ is isolated as a precursor ion in the quadrupole mass analyzer and fragmented by collision-induced dissociation in a collision cell. The fragment ions are then recorded in the Orbitrap mass analyzer and specific ions for $A\beta_{1\text{-}42}$ are then selected to measure the abundance of the peptide. Quantification can be performed by adding a known amount of a stable isotope labeled (e.g., 13C) homolog of the peptide prior to sample preparation.

For this LC-MS method, we used the surrogate analyte approach [11–14] which enables calibration in human CSF, instead of using a surrogate matrix. For this surrogate analyte approach, two different isotopically labeled standards are used. One ($^{15}N\text{-}A\beta_{1\text{-}42}$) is used to generate the calibration curve in human CSF while another ($^{13}C\text{-}A\beta_{1\text{-}42}$) is used as internal standard. The concentration of endogenous $A\beta_{1\text{-}42}$ in unknown samples is then calculated using the fit of the calibration curve constructed using the $^{15}N\text{-}A\beta_{1\text{-}42}/^{13}C\text{-}A\beta_{1\text{-}42}$ ratio by the calculated endogenous $A\beta_{1\text{-}42}/^{13}C\text{-}A\beta_{1\text{-}42}$ ratio measured in unknown samples. The surrogate analyte approach was chosen since there is no analyte-free CSF available, and low $A\beta_{1\text{-}42}$ recovery was observed when using unlabeled $A\beta_{1\text{-}42}$ in a surrogate matrix, such as artificial CSF, during method development.

2 Materials

The materials used in this chapter is listed in Table 4.1. For safety information see Figure 4.2.

Table 4.1
Materials

Name	Company	Catalog number
Thermo Scientific Q-exactive Hybrid Quadrupole Orbitrap Mass Spectrometer	Thermo Fisher Scientific	IQLAAEGAA-PFALGMAZR
UltiMate™ TCC-3000RS Rapid Separation Thermostatted Column Compartment	Thermo Fisher Scientific	5730.0000
UltiMate™ WPS-3000RS/TRS Rapid Separation Well Plate Autosampler	Thermo Fisher Scientific	5840.0020
UltiMate™ HPG-3400RS Rapid Separation Binary Pump	Thermo Fisher Scientific	5040.0046
UltiMate™ HPG-3200SD Standard Binary Pump	Thermo Fisher Scientific	5040.0021
SRD-3x00 Solvent Racks with Degassers for UltiMate™ 3000 Pumps	Thermo Fisher Scientific	5035.9230
Column Switching Valve for UltiMate™ 3000 RSLC Nano Systems	Thermo Fisher Scientific	6041.0001
Savant SC210A SpeedVac Concentrator	Thermo Fisher Scientific	SC210A-230
Microplate shakers, TiMix 2	Edmund Bühler	6110 000
Synergy® UV Water Purification System	Merck	SYNSVHFWW
Extraction Plate Manifold for Oasis 96-Well Plates	Waters	186001831
Ammonium hydroxide solution, puriss. p.a., reag. ISO, reag. Ph. Eur., ~25% NH3 basis	Sigma-Aldrich	30501-1L-M
Acetonitrile, Far UV HPLC Gradient grade, 2.5 L	Fisher Scientific	A/0627/17X
Ortho-Phosphoric acid 85%, ACS,ISO,Reag. Ph Eur	Merck Millipore	1005731000
Guanidine- Hydrochloride, 500 g	Thermo Fisher Scientific	24110
Bovine Serum Albumin, lyophilized powder, suitable for (for molecular biology), Non-acetylated. 25 mg	Sigma-Aldrich	B6917-25MG
Beta-Amyloid (1-42), Ultra Pure, HFIP	rPeptide	A-1163-2
15N Beta-Amyloid (1-42), Uniformly labeled	rPeptide	A-1102-2
13C Beta-Amyloid (1-42), Uniformly labeled	rPeptide	A-1106-2
Protein LoBind, 0.5 mL, PCR clean, colorless, 100 tubes (2 bags × 50 tubes)	Eppendorf	0030108094

(continued)

Table 4.1 (continued)

Name	Company	Catalog number
Protein LoBind, 1.5 mL, PCR clean, colorless, 100 tubes (2 bags × 50 tubes)	Eppendorf	0030108116
Protein LoBind, 2.0 mL, PCR clean, colorless, 100 tubes (2 bags × 50 tubes)	Eppendorf	0030108132
Protein LoBind, 5.0 mL, PCR clean, colorless, 100 tubes (2 bags × 50 tubes)	Eppendorf	0030108302
Protein LoBind, wells colorless, 1000 μL, PCR clean, white, 20 plates (5 bags × 4 plates)	Eppendorf	0030504208
0.75 mL Non coded Screw Cap tubes V-bottom Snap Bulk	Micronic	MP32069L
Split TPE Capcluster for capping 96 individual tubes—VARIA package: Gray, Natural, Yellow, Red, Pink, Blue, Light Green, Black (10 TPE Capclusters per color)	Micronic	MP53075
Loborack-96 White, Laser etched (including low cover) (Barcoded A1-H1 side)	Micronic	MPW51013BC3
Optifit Tip, 50–1200 μL, extended, single tray	Sartorius	791210
Optifit Tip, 50–1200 μL, extended, refill pack	Sartorius	791212
Oasis MCX 96-well μElution Plate, 2 mg Sorbent per Well, 30 μm Particle Size, 1/pk	Waters	186001830BA
Plate manifold reservoir tray 25/pkg	Waters	WAT058942
Dionex ProSwift RP-4H Monolith Column (1.0 × 250 mm)	Fisher Scientific	066640

3 Methods

3.1 Preparation of Solutions

1. Prepare 100 mL peptide diluent consisting of 20% (v/v) acetonitrile (ACN) and 4% (v/v) concentrated ammonia solution in ultra-pure water at 18 MΩ resistivity. Add 4 mL concentrated ammonia (~25%) and 20 mL ACN to 76 mL ultra-pure water and mix gently. Make fresh daily.
2. Prepare 50 mL of 5 M guanidine-hydrochloride by dissolving 26.08 g guanidine-hydrochloride in deionized water to a final volume of 50 mL. Store at 20 °C and make fresh monthly.
3. Prepare 200 mL of ~4% phosphoric acid in deionized water (v/v) by adding 9.4 mL concentrated phosphoric acid (~85%) to 190 mL water. Adjust the final volume to 200 mL with deionized water. Store in the refrigerator and make fresh weekly.
4. Prepare 100 mL of 75% ACN and 10% concentrated ammonia solution (v/v) in deionized water by adding 75 mL ACN and 10 mL concentrated ammonia (~25%) to 15 mL ultra-pure water. Make fresh daily.
5. Thaw at least 2.5 mL human CSF (for use in the calibrators) (obtained from de-identified leftover samples from clinical routine analysis).
6. Prepare artificial CSF containing 8.66 g/L NaCl, 0.224 g/L KCl, 0.206 g/L $CaCl_2$ $2H_2O$, 0.163 g/L $MgCl_2$ $6H_2O$, 0.214 g/L Na_2HPO_4 $7H_2O$ and 0.027 g/L NaH_2PO_4 H_2O in ultra-pure water (electrolyte concentrations 150 mM Na, 3.0 mM K, 1.4 mM Ca, 0.8 mM Mg, 1.0 mM P, and 155 mM Cl) and add bovine serum albumin to a final concentration of 4 g/L. Only 1 mL is needed per analysis but prepare a large volume, aliquot, and store for future use.

3.2 Preparation of Calibrators

This protocol requires aliquots of at least 50 µL with a concentration of 50 µg/mL for each Aβ peptide as starting material. The Aβ peptides should be dissolved in peptide diluent (20% ACN and 4% concentrated ammonia in ultra-pure water, prepared in Subheading 3.1.1) and stored at −80 °C.

1. Prepare 0.5 mL 4 µg/mL $^{15}N\text{-}A\beta_{1\text{-}42}$ peptide by adding 40 µL of 50 µg/mL $^{15}N\text{-}A\beta_{1\text{-}42}$ to 0.46 mL peptide diluent in a 0.5 mL microcentrifuge tube. Mix on vortex mixer for 1 min.
2. Prepare 2 mL 100 ng/mL $^{15}N\text{-}A\beta_{1\text{-}42}$ peptide by adding 50 µL of the 4 µg/mL $^{15}N\text{-}A\beta_{1\text{-}42}$ to 1.95 mL peptide diluent in a 2 mL microcentrifuge tube. Mix on vortex mixer for 1 min.
3. Prepare six calibrator solutions (A-F) by mixing the volumes of each solution indicated in Table 4.2. Use 0.5, 1.5, and 2 mL microcentrifuge tubes. Mix on vortex mixer for 1 min.

Table 4.2
Calibrator solutions. Calibrator solutions prepared in 20% ACN and 4% concentrated ammonia used for spiking CSF calibrators

Calibrator solution	Volume of 100 ng/mL ^{15}N-Aβ_{1-42} solution (mL)	Volume of 20% ACN and 4% ammonia (mL)	Final volume (mL)	Final ^{15}N-Aβ_{1-42} concentration (ng/mL)
A	0.20	0.30	0.50	40.00
B	0.15	0.35	0.50	30.00
C	0.20	0.80	1.00	20.00
D	0.10	0.90	1.00	10.00
E	0.05	0.95	1.00	5.00
F	0.03	1.97	2.00	1.50

Table 4.3
Calibrators. Calibrators prepared in human CSF

Calibrator	Volume of corresponding calibrator solution (mL)	Volume of human CSF (mL)	Final volume (mL)	Final 15N-Aβ1-42 concentration (ng/mL)
A	0.02 (A)	0.18	0.20	4.00
B	0.02 (B)	0.18	0.20	3.00
C	0.02 (C)	0.18	0.20	2.00
D	0.02 (D)	0.18	0.20	1.00
E	0.02 (E)	0.18	0.20	0.50
F	0.02 (F)	0.18	0.20	0.15

4. Prepare the final calibrators (in duplicate) in 0.5 mL microcentrifuge tubes by adding corresponding calibration solutions and human CSF according to Table 4.3. Mix on vortex mixer for 1 min.

3.3 Preparation of Internal Standard

1. Prepare 2 mL 0.8 μg/mL ^{13}C-Aβ_{1-42} peptide by adding 32 μL of 50 μg/mL ^{13}C-Aβ_{1-42} to 1.968 mL peptide diluent in a 2 mL microcentrifuge tube. Mix on vortex mixer for 1 min.
2. Prepare 5 mL 16 ng/mL ^{13}C-Aβ_{1-42} peptide by adding 0.1 mL of 0.8 μg/mL to 4.9 mL peptide diluent in a 5 mL microcentrifuge tube. Mix on vortex mixer for 1 min.

3.4 Preparation of Response Factor Sample

Note: The response factor (RF) determination is performed to determine the concentration of the labeled peptide used for calibration (^{15}N-$A\beta_{1-42}$). This requires that the concentration of the unlabeled $A\beta_{1-42}$ peptide has been determined using amino acid analysis (AAA).

1. Prepare 0.5 mL 4 μg/mL unlabeled $A\beta_{1-42}$ by adding 40 μL of 50 μg/mL unlabeled $A\beta_{1-42}$ to 0.46 mL peptide diluent in a 0.5 mL microcentrifuge tube. Mix on vortex mixer for 1 min.
2. Prepare a 2 mL 40 ng/mL mix of unlabeled and ^{15}N-$A\beta_{1-42}$ by adding 20 μL of 4 μg/mL unlabeled $A\beta_{1-42}$ and 20 μL of 4 μg/mL ^{15}N-$A\beta_{1-42}$ to 1.96 mL of peptide diluent in a 2 mL microcentrifuge tube. Mix on vortex mixer for 1 min.
3. Add 20 μL of the 40 ng/mL mix to 0.38 mL artificial CSF in a 0.5 mL microcentrifuge tube. Prepare duplicates and mix on vortex mixer for 1 min.

3.5 Sample Preparation

3.5.1 Note: Thaw Samples to Be Measured at Room Temperature on a Roller

1. Add 0.18 mL of each calibrator, response factor and unknown sample (including quality control [QC] samples if used) to a 1 mL protein 96 deep-well plate according to Fig. 4.1 (assuming a full plate is used). Make sure to add the samples in, or close to the bottom of the wells.
2. Add 20 μL of internal standard to each well (i.e., calibrators, response factors, QCs, and unknowns). It is crucial to release the drop on the side of the well close to the surface of the sample without submerging the pipette tip.
3. Add 0.2 mL 5 M guanidine-hydrochloride to each well.
4. Place the sample plate on a microplate shaker and mix the samples for 45 min at 1100 rpm. The optimal frequency might differ depending on instrumentation. Set the frequency

Calibrators	Unknowns	Response factor	Qality control

	1	2	3	4	5	6	7	8	9	10	11	12
A	A						A					
B	B						B					
C	C						C					
D	D						D					
E	E						E					
F	F						F					
G	RF						RF					
H	QC						QC					

Fig. 4.1 SPE and deep-well plates layout. Typical layout of calibrators (**a–f**), response factor sample (RF), quality control samples (QC), and unknowns

and amplitude of the mixer so that the solutions are thoroughly mixed and no drops of internal standard or CSF are left unmixed on the side of the wells.

5. Add 0.2 mL of 4% phosphoric acid to each well. Vortex mix briefly.

3.6 Solid Phase Extraction

Note: In all washing, loading and elution steps, apply lowest possible vacuum after adding the solution and gradually increase as needed to load or elute the solution. Disable the vacuum between each loading and elution step.

1. Put a reservoir tray for waste under a mixed-mode cation exchange 96-well solid phase extraction (SPE) plate in the extraction plate manifold chamber.
2. Condition the SPE sorbent by adding 0.2 mL methanol to each well.
3. Equilibrate the sorbent by adding 0.2 mL 4% phosphoric acid to each well.
4. Transfer all samples (about 0.62 mL in each well) from the deep-well plate to the SPE-plate. It is highly recommended to use an eight channel pipette when transferring the samples from the deep well-plate to the SPE-plate. It is not crucial to transfer the entire or equal volumes of all samples since the samples contain an internal standard which will compensate for variations.
5. Wash the sorbent after the samples have passed through by adding 0.2 mL 4% phosphoric acid to each well.
6. After the washing solvent has eluted from the sorbent, replace the reservoir tray with a collection plate or tubes.
7. Elute the sample from the sorbent twice by adding 50 μL 75% ACN and 10% concentrated ammonia solution in ultra-pure water. Note that this solution requires very low vacuum to pass through the sorbent. Remember to disable the vacuum between each addition.
8. OPTIONAL. Seal the collection plate or tubes and freeze at −80 °C and remove the seal from the collection plate or tubes before proceeding to **step 9**.
9. Dry the eluates by using vacuum centrifugation (without applying heat). This can take from one to several hours depending on the vacuum centrifuge.
10. Seal the containers and freeze at −80 °C.

3.7 Liquid Chromatography

Prepare mobile phase A (5% ACN and 0.3% concentrated ammonia in deionized water [v/v]), B (4% deionized water and 0.1% concentrated ammonia in ACN [v/v]) and needle wash (50% ACN and 4% concentrated ammonia in deionized water [v/v]).

1. For 500 mL mobile phase A, add 25 mL ACN and 1.5 mL concentrated ammonia to 475 mL ultra-pure water.
2. For 500 mL mobile phase B, add 500 μL concentrated ammonia and 25 mL ultra-pure water to 475 mL ACN.
3. Prepare 250 mL needle wash by adding 120 mL ACN and 10 mL concentrated ammonia to 120 mL ultra-pure water. Adjust the final volume to 250 mL with ultra-pure water.
4. Put mobile phase A, B, and needle wash bottles open in sonication bath for 20 min before use with the LC system.
5. Dissolve each sample with 25 μL peptide diluent and place on shaker for 20 min. Centrifuge down the sample and place in the autosampler (keep at 7 °C).
6. Inject 20 μL sample on a 1 × 250 mm polystyrene-divinylbenzene (reversed-phase) monolithic column maintained at 50 °C.
7. Use the LC gradient shown in the Table 4.4 with a flow rate of 0.3 mL/min. Divert the first two and last 5 min to waste (post column) using a divert valve to reduce contamination of the mass spectrometer.

3.8 Mass Spectrometric Analysis

Note: These parameters were used for a quadrupole-Orbitrap hybrid mass spectrometer equipped with a heated electrospray ionization source.

1. Set the parameters for the ion source according to Table 4.5.
2. Set the MS instrument to isolate the 4+ charge states of unlabeled $A\beta_{1\text{-}42}$ (1129.48 mass-to-charge ratio [*m/z]*), $^{15}N\text{-}A\beta_{1\text{-}42}$ (1143.00 *m/z*) and $^{13}C\text{-}A\beta_{1\text{-}42}$ (1179.50 *m/z*) in the quadrupole mass analyzer with an isolation width of 2.5 *m/z*.
3. Fragment the isolated peptides in the collision cell with a normalized collision energy (NCE) of 17.0. This might need

Table 4.4
LC gradient. The LC gradient used with a constant flow rate of 300 μL/min

Time (min)	% mobile phase B
0	5
1	5
6	20
7	90
9	90
10	5
15	5

Table 4.5
Ion source settings. Parameters for the ion source to be set in the instrument tune software

Parameter	Value
Sheath gas	50
Auxiliary gas	6
Spray voltage	4.4 kV
S-lens RF	61
Heater temperature	+190 °C
Capillary temperature	+350 °C

to be tuned for each instrument even of the same type (and especially if using other types of instrument, e.g., a triple quadrupole MS).

4. Record fragment spectra with a resolution of 17.500 with an automatic gain control target of 2×10^5 charges and a maximum injection time of 250 ms.

3.9 Data Processing

1. Use the sum of the product ions (with a mass tolerance of ±250 milli mass units [mmu]) in Table 4.6 to calculate the chromatographic areas for each peptide. Note the ion types (b ion) and charge states are only shown for unlabeled $A\beta_{1\text{-}42}$ product ions. Both $^{15}N\text{-}A\beta_{1\text{-}42}$ and $^{13}C\text{-}A\beta_{1\text{-}42}$ fragments m/z values are put in the same order as the unlabeled $A\beta_{1\text{-}42}$ fragments.
2. Determine the average response factor of the two response factor samples by dividing the area under the curve (chromatographic peak) of $^{15}N\text{-}A\beta_{1\text{-}42}$ with the area under the curve of unlabeled $A\beta_{1\text{-}42}$.
3. Adjust the concentration of the $^{15}N\text{-}A\beta_{1\text{-}42}$ used for calibration by multiplying it with the response factor calculated in Subheading 3.9.2.
4. Construct a calibration curve by plotting the area ratios of the response factor corrected $^{15}N\text{-}A\beta_{1\text{-}42}$ to the internal standard ($^{13}C\text{-}A\beta_{1\text{-}42}$) from the two sets of calibrators against the concentration (Fig. 4.2).
5. Calculate the slope and intercept of the calibration curve using linear regression with 1/x weighting.
6. Calculate the area ratio of unlabeled $A\beta_{1\text{-}42}$ to the internal standard ($^{13}C\text{-}A\beta_{1\text{-}42}$) for unknown samples. A representative chromatogram for unknown samples can be seen in Fig. 4.3.

Table 4.6
Ions used for quantification. The 4+ charge states of the precursor ions are isolated in the quadrupole mass analyzer with an isolation width of 2.5 *m/z*. The product ions (with a mass tolerance of ±250 mmu) are used to calculate the chromatographic areas for each peptide. Ion types and numbers are only shown for unlabeled $A\beta_{1-42}$ product ions since they are the same for both ^{15}N-$A\beta_{1-42}$ and ^{13}C-$A\beta_{1-42}$

Precursor ion	Product ions
Unlabeled $A\beta_{1-42}$ (m/z 1129.58, 4+)	915.19 ($b33^{4+}$), 943.21 ($b34^{4+}$), 975.98 ($b35^{4+}$), 1000.74 ($b36^{4+}$), 1029.51 ($b38^{4+}$), 1054.03 ($b39^{4+}$), 1078.79 ($b40^{4+}$), 1107.06 ($b41^{4+}$), 1163.23 ($b31^{3+}$), 1200.25 ($b32^{3+}$), 1257.29 ($b34^{3+}$), 1300.96 ($b35^{3+}$), 1333.66 ($b36^{3+}$), 1372.00 ($b38^{3+}$), 1405.02 ($b39^{3+}$)
^{15}N-$A\beta_{1-42}$ (m/z 1143.00, 4+)	926.41, 954.68, 987.95, 1012.71, 1041.22, 1066.99, 1091.75, 1120.28, 1177.18, 1215.55, 1272.58, 1316.92, 1349.94, 1388.63, 1422.31
^{13}C-$A\beta_{1-42}$ (m/z 1179.50, 4+)	955.33, 985.11, 1019.37, 1045.14, 1074.65, 1100.67, 1126.69, 1156.40, 1253.43, 1313.14, 1358.50, 1393.19, 1432.21, 1466.90

Chemical	CLP	Measure
2-propanol (isopropanol)		Use gloves
Guanidine-HCl		Use gloves
Methanol		Use nitrile gloves that can withstand methanol.
Acetonitrile		Use nitrile gloves that can withstand acetonitrile.
Phosphoric acid ~85%		Use gloves
Ammonium hydroxide ~25%		Use gloves. Work in fume hood.

Fig. 4.2 Safety information. Safety information for chemicals used for this protocol

7. Interpolate the concentration of unknown samples from the calibration curve (Fig. 4.4) using the slope and intercept obtained in **step 5**.

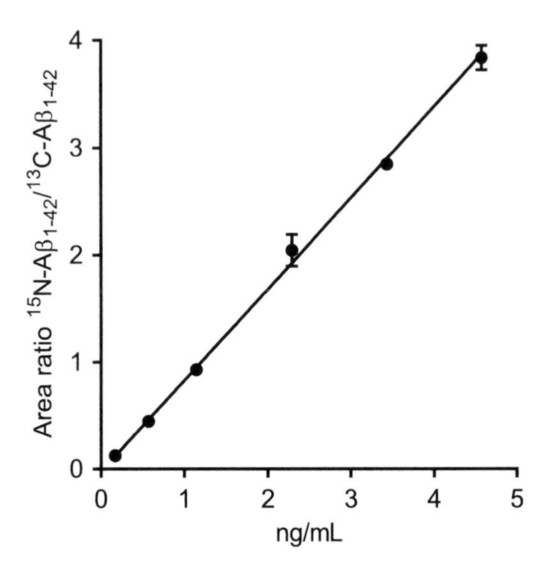

Fig. 4.3 Calibration curve. Calibration curve constructed using ^{15}N-$A\beta_{1-42}$ at 172, 572, 1144, 2287, 3431, and 4574 pg/mL (adjusted using the response factor) and ^{13}C-$A\beta_{1-42}$ as internal standard in human CSF ($n = 2$). The area ratio of ^{15}N-$A\beta_{1-42}$/^{13}C-$A\beta_{1-42}$ is plotted (Y-axis) against the concentration (X-axis)

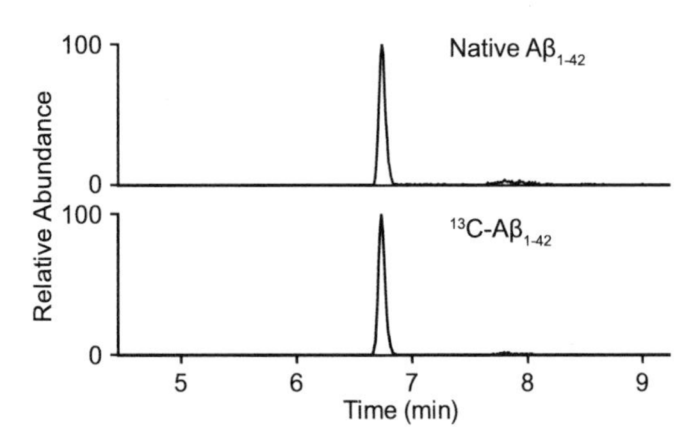

Fig. 4.4 Chromatogram. Chromatogram of 0.500 ng/mL unlabeled (endogenous) $A\beta_{1-42}$ (top panel) and 1.6 ng/mL ^{13}C-$A\beta_{1-42}$ (bottom panel) in human CSF

4 Notes

Since ^{15}N-$A\beta_{1-42}$ and unlabeled $A\beta_{1-42}$ may give different responses in the mass spectrometer, the concentration of ^{15}N-$A\beta_{1-42}$ is adjusted by measuring a response factor (RF) sample—an artificial CSF sample containing equal concentrations of ^{15}N-$A\beta_{1-42}$ and unlabeled $A\beta_{1-42}$ with known concentration determined by AAA. The RF might differ due to possible variations in the isotopic purity of the ^{15}N-labeled peptide between batches, but also different mass spectrometers and day-to-day methodological variation can affect the RF. Therefore, the response factor should be determined for each measurement day.

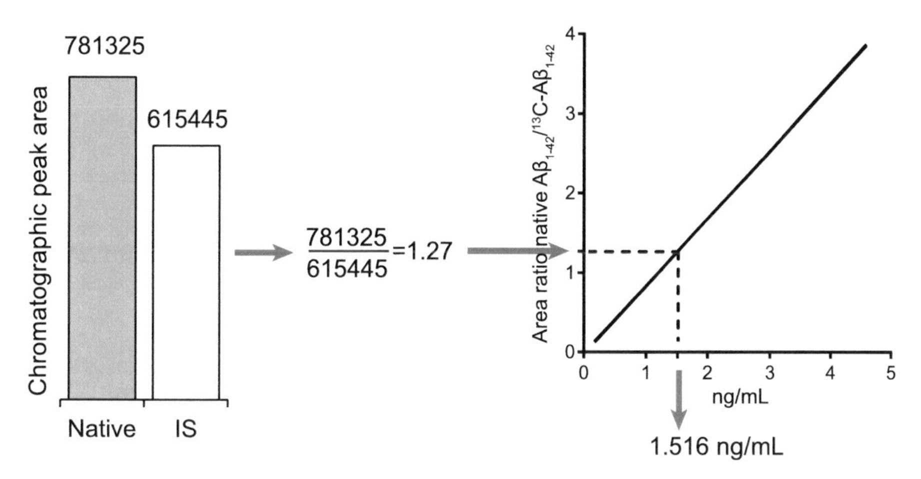

Fig. 4.5 Quantification of unknown $A\beta_{1-42}$ in unknown samples. The peak area ratio is calculated by dividing the unlabeled $A\beta_{1-42}$ chromatographic peak area with the internal standard (^{13}C-$A\beta_{1-42}$) chromatographic peak area. The concentration of unlabeled $A\beta_{1-42}$ in the sample is interpolated from the calibration curve

The most critical steps in this protocol are the preparation of calibrators and RF samples. The $A\beta_{1-42}$ peptide is very hydrophobic and easily stick to pipette tips and surfaces of tubes [6, 15, 16]. To minimize loss of Aβ peptides during pipetting it is very important to saturate the pipette tips prior to delivery. Preferably, three volumes of peptide solution should be discarded prior to delivery to a new tube containing solution, instead of saturating the tip by pipetting up and down in the original tube. Depending on the volume and concentration of the stock solution this is not always possible. The second best approach is of course to pipette the peptide solution up and down three times prior to delivery. For the same reason, it is important to use appropriate sizes for tubes, avoiding large void volumes.

One obvious limitation of this technique is its low throughput compared to automated immunoassays. One full 96-well plate containing 12 calibrators, two response factor samples, two QC samples and 80 unknown samples is typically processed per day, while two plates can be processed in sequence by an experienced operator. However, the purpose of the described method is high accuracy and not throughput. Another limitation of this method is that the operator will need extensive mass spectrometry training before running the analysis on the instrument.

References

1. Winblad B, Amouyel P, Andrieu S, Ballard C, Brayne C, Brodaty H, Cedazo-Minguez A, Dubois B, Edvardsson D, Feldman H, Fratiglioni L, Frisoni GB, Gauthier S, Georges J, Graff C, Iqbal K, Jessen F, Johansson G, Jonsson L, Kivipelto M, Knapp M, Mangialasche F, Melis R, Nordberg A, Rikkert MO, Qiu C, Sakmar TP, Scheltens P, Schneider LS, Sperling R, Tjernberg LO, Waldemar G, Wimo A, Zetterberg H

(2016) Defeating Alzheimer's disease and other dementias: a priority for European science and society. Lancet Neurol 15 (5):455–532. https://doi.org/10.1016/S1474-4422(16)00062-4

2. Masters CL, Bateman R, Blennow K, Rowe CC, Sperling RA, Cummings JL (2015) Alzheimer's disease. Nat Rev Dis Primers 1:15056. https://doi.org/10.1038/nrdp.2015.56
3. Blennow K, Mattsson N, Scholl M, Hansson O, Zetterberg H (2015) Amyloid biomarkers in Alzheimer's disease. Trends Pharmacol Sci 36(5):297–309. https://doi.org/10.1016/j.tips.2015.03.002
4. Blennow K, Hampel H, Weiner M, Zetterberg H (2010) Cerebrospinal fluid and plasma biomarkers in Alzheimer disease. Nat Rev Neurol 6(3):131–144. https://doi.org/10.1038/nrneurol.2010.4
5. Dubois B, Feldman HH, Jacova C, Hampel H, Molinuevo JL, Blennow K, DeKosky ST, Gauthier S, Selkoe D, Bateman R, Cappa S, Crutch S, Engelborghs S, Frisoni GB, Fox NC, Galasko D, Habert MO, Jicha GA, Nordberg A, Pasquier F, Rabinovici G, Robert P, Rowe C, Salloway S, Sarazin M, Epelbaum S, de Souza LC, Vellas B, Visser PJ, Schneider L, Stern Y, Scheltens P, Cummings JL (2014) Advancing research diagnostic criteria for Alzheimer's disease: the IWG-2 criteria. Lancet Neurol 13(6):614–629. https://doi.org/10.1016/S1474-4422(14)70090-0
6. Bjerke M, Portelius E, Minthon L, Wallin A, Anckarsater H, Anckarsater R, Andreasen N, Zetterberg H, Andreasson U, Blennow K (2010) Confounding factors influencing amyloid Beta concentration in cerebrospinal fluid. Int J Alzheimers Dis 2010:986310. https://doi.org/10.4061/2010/986310
7. Dumurgier J, Vercruysse O, Paquet C, Bombois S, Chaulet C, Laplanche JL, Peoc'h K, Schraen S, Pasquier F, Touchon J, Hugon J, Lehmann S, Gabelle A (2013) Intersite variability of CSF Alzheimer's disease biomarkers in clinical setting. Alzheimers Dement 9(4):406–413. https://doi.org/10.1016/j.jalz.2012.06.006
8. Mattsson N, Andreasson U, Persson S, Carrillo MC, Collins S, Chalbot S, Cutler N, Dufour-Rainfray D, Fagan AM, Heegaard NH, Robin Hsiung GY, Hyman B, Iqbal K, Lachno DR, Lleo A, Lewczuk P, Molinuevo JL, Parchi P, Regeniter A, Rissman R, Rosenmann H, Sancesario G, Schroder J, Shaw LM, Teunissen CE, Trojanowski JQ, Vanderstichele H, Vandijck M, Verbeek MM, Zetterberg H, Blennow K, Kaser SA, Alzheimer's Association QCPWG (2013) CSF biomarker variability in the Alzheimer's association quality control program. Alzheimers Dement 9(3):251–261. https://doi.org/10.1016/j.jalz.2013.01.010
9. Kang JH, Korecka M, Toledo JB, Trojanowski JQ, Shaw LM (2013) Clinical utility and analytical challenges in measurement of cerebrospinal fluid amyloid-beta1-42 and tau proteins as Alzheimer disease biomarkers. Clin Chem 59(6):903–916. https://doi.org/10.1373/clinchem.2013.202937
10. Mattsson N, Zegers I, Andreasson U, Bjerke M, Blankenstein MA, Bowser R, Carrillo MC, Gobom J, Heath T, Jenkins R, Jeromin A, Kaplow J, Kidd D, Laterza OF, Lockhart A, Lunn MP, Martone RL, Mills K, Pannee J, Ratcliffe M, Shaw LM, Simon AJ, Soares H, Teunissen CE, Verbeek MM, Umek RM, Vanderstichele H, Zetterberg H, Blennow K, Portelius E (2012) Reference measurement procedures for Alzheimer's disease cerebrospinal fluid biomarkers: definitions and approaches with focus on amyloid beta42. Biomark Med 6(4):409–417. https://doi.org/10.2217/bmm.12.39
11. Leinenbach A, Pannee J, Dulffer T, Huber A, Bittner T, Andreasson U, Gobom J, Zetterberg H, Kobold U, Portelius E, Blennow K, IFCC Scientific Division Working Group on CSF Proteins (2014) Mass spectrometry-based candidate reference measurement procedure for quantification of amyloid-beta in cerebrospinal fluid. Clin Chem 60(7):987–994. https://doi.org/10.1373/clinchem.2013.220392
12. Ahmadkhaniha R, Shafiee A, Rastkari N, Kobarfard F (2009) Accurate quantification of endogenous androgenic steroids in cattle's meat by gas chromatography mass spectrometry using a surrogate analyte approach. Anal Chim Acta 631(1):80–86. https://doi.org/10.1016/j.aca.2008.10.011
13. Jemal M, Schuster A, Whigan DB (2003) Liquid chromatography/tandem mass spectrometry methods for quantitation of mevalonic acid in human plasma and urine: method validation, demonstration of using a surrogate analyte, and demonstration of unacceptable matrix effect in spite of use of a stable isotope analog internal standard. Rapid Commun Mass Spectrometry 17(15):1723–1734. https://doi.org/10.1002/rcm.1112
14. Li W, Cohen LH (2003) Quantitation of endogenous analytes in biofluid without a true blank matrix. Anal Chem 75 (21):5854–5859. https://doi.org/10.1021/ac034505u

15. Perret-Liaudet A, Pelpel M, Tholance Y, Dumont B, Vanderstichele H, Zorzi W, ElMoualij B, Schraen S, Moreaud O, Gabelle A, Thouvenot E, Thomas-Anterion C, Touchon J, Krolak-Salmon P, Kovacs GG, Coudreuse A, Quadrio I, Lehmann S (2012) Cerebrospinal fluid collection tubes: a critical issue for Alzheimer disease diagnosis. Clin Chem 58(4):787–789. https://doi.org/10.1373/clinchem.2011.178368

16. Lewczuk P, Beck G, Esselmann H, Bruckmoser R, Zimmermann R, Fiszer M, Bibl M, Maler JM, Kornhuber J, Wiltfang J (2006) Effect of sample collection tubes on cerebrospinal fluid concentrations of tau proteins and amyloid beta peptides. Clin Chem 52 (2):332–334. https://doi.org/10.1373/clinchem.2005.058776

Chapter 5

Immunoassay and Mass Spectrometry Methods for Tau Protein Quantification in the Cerebrospinal Fluid

Christophe Hirtz, Jérôme Vialaret, Constance Delaby, Aleksandra Maleska Maceski, Cyndi Catteau, Nelly Ginestet, Audrey Gabelle, and Sylvain Lehmann

Abstract

Quantification of tau proteins in the cerebrospinal fluid is a major biomarker for the positive diagnosis of Alzheimer's disease, as well as for other neurodegenerative diseases and brain damage monitoring. Tau is present in this fluid in multiple proteoforms, different by their lengths and phosphorylation sites. Immunodetection approaches for "total" tau and its phosphorylation form pTau(181) are now being used in routine with IVD qualified kits represented by classical ELISA and fully automatized systems. There is however a need for a reference method represented by mass spectrometry which could help to homogenize results across platforms and help to decipher the diversity of the tau isoforms in CSF.

Key words Tau proteins, CSF, ELISA, Automatized system, Mass spectrometry

1 Introduction

Tau protein, which is mainly expressed in the central nervous system, is a microtubule-associated protein which role is to stabilize tubulin polymerization, a mechanism essential for efficient intracellular transport and axonal growth [1]. Tau exists in six different isoforms and the regulation of its function is ensured by its phosphorylation over about 80 putative sites scattered within the entire protein [2] and notably at position 181 (pTau(181)). In Alzheimer's disease (AD), its hyperphosphorylation will promote its aggregation and the formation of neurofibrillary tangles in the brain, which are one of the histological hallmarks of the disease [2]. Incidentally, Tau is released in the cerebrospinal fluid (CSF) following cell damages and through secretion pathways which still need to be fully characterized. There is an important variety of Tau protein isoforms different by their length and phosphorylation sites present in this fluid [3]. Importantly, AD is associated with an

Charlotte E. Teunissen and Henrik Zetterberg (eds.), *Cerebrospinal Fluid Biomarkers*, Neuromethods, vol. 168,
https://doi.org/10.1007/978-1-0716-1319-1_5,

increase in Tau and pTau(181) levels in this fluid [4]. These two biomarkers, along with the detection of amyloid peptides are now included in the diagnosis criteria for the disease [5]. The use of these biomarkers increases the diagnostic precision in case of atypical forms and refine differential diagnoses; furthermore, these biomarkers seem to have a predictive value on the cognitive decline and disease prognosis [6]. Their clinical use is now implemented in many clinical laboratories [7] which use IVD qualified kits that are conformed with the requirement of the clinical norm ISO15189.

Here, we first describe the implementation of clinical CSF tau assays in their classical ELISA form, as well as in their automatized, random access format. The description of the handling of the samples in their pre-analytical and post-analytical steps are generic for all the assays present on the IVD market. We then provide the detailed method for the quantitative mass detection of CSF tau based on the antibody-free multiple reaction monitoring of a non-phosphorylated peptide of its central core region [8]. This method, along with the generation of a tau-certified reference material, could help in the future homogenize results of IVD assays.

2 Materials

All solutions used in this study were prepared with ultrapure water with a resistivity of 18 MO.cm at 25 °C.

2.1 ELISA Tau

1. EUROIMMUN Analyzer I-2P, automated ELISA processing—for up to 3 microplates.
2. tTau Euroimmun IVD kit (ref: EQ6531-9601L).
3. ?Tau Euroimmun IVD kit (ref: EQ6591-9601L).
4. Pointe Diamond D200 and D1000.

2.2 CLEIA Tau

1. Lumipulse, CLEIA—Chemiluminescent Enzyme Immunoassay analyzer, G600i.
2. Lumipulse G Total Tau Immunoreaction Cartridges (IRC), Art. no. 230312 (3 × 14 Tests—CE).
3. Lumipulse G Total Tau Calibrators set, Art. no. 230329 (2 × 3 concentrations—CE).
4. Lumipulse G pTau 181 Immunoreaction Cartridges (IRC), KEY-CODE: FRI29776 (3 × 14 Tests).
5. Lumipulse G pTau 181 Calibrators set, KEY-CODE: FRI41969 (2 × 3 concentrations).

2.3 Mass Spectrometry

LC-MS system, 1290 Infinity LC—6490 triple Quad (Agilent), with control computer with MassHunter software (Agilent).

2.3.1 Tau Standard Preparation

1. Buffered solution of 14N and 15N-tau-441 recombinant protein (Guy Lippens, UMR 8525, Lille Pasteur Institute). Concentration of 14N and 15N tau primary calibrators were determined by amino acid analysis. Molecular weights (MWs) of 14N-tau-441 recombinant proteins were confirmed by MALDI-TOF MS. Isotopic incorporation of 15N in recombinant protein was estimated at ~99% based on the isotopic profiles of tryptic peptides observed by LC-ESI-HRMS.
2. 50 mmol/L ammonium bicarbonate: resuspend 39.54 mg of ammonium bicarbonate with 10 mL of water. Vortex few seconds.
3. 1 mmol/L BSA: Dissolve 5 g of Bovine Serum Albumin (BSA) with 76 mL of water to obtain a solution at 1 mmol/L. Vortex at minimal speed for 30 min.
4. 0.5% goat serum: Dilute 2 μL of pure Normal Goat Serum in 198 μL of water.

2.3.2 Protein Precipitation

1. 70% perchloric acid.

2.3.3 Solid Phase Extraction

This step must be performed under a fume hood because of the organic and acidic vapors release.

1. Collection plate: a Deepwell Plate 96/2000 μL with clear wells, 2000 μL, PCR clean, white.
2. Elution plate: an Eppendorf twin.tec PCR Plate 96 LoBind, skirted, PCR clean, colorless (Eppendorf).
3. Solvent A: 100% methanol ULC-MS grade.
4. Solvent B: 0.1% trifluoroacetic acid (TFA). To obtain solvent B, mix 9 mL of ultrapure water and 1 mL of 10% TFA.
5. =99.0% TFA LC-MS grade.
6. Dilute TFA giving a 10% solution by mixing 9 mL of ultrapure water and 1 mL of TFA =99.0% (**Note 1**).
7. Solvent C: 35% acetonitrile/65% water with 0.1% TFA. Acetonitrile solvent is ULC-MS grade. To obtain solvent C, 3.5 mL of acetonitrile is mixed with 6.5 mL of 0.1% TFA.
8. Oasis HLB 96-well μElution Plate, 2 mg Sorbent per Well, 30 μm Particle Size (Waters).

2.3.4 Sample Drying

Equilibration plate: an Eppendorf twin.tec PCR Plate 96 LoBind, skirted, PCR clean, colorless.

2.3.5 Protein Digestion

1. 50 mmol/L ammonium bicarbonate: Resuspend 39.54 mg of ammonium bicarbonate with 10 mL of water. Vortex few seconds.
2. Resuspend 100 μg of lyophilized trypsin Gold (Mass Spec Grade, Promega) at 1 μg/μL with 50 mmol/L ammonium bicarbonate (**Note 2**). Vortex on ice 1 min at 1000 rpm.
3. Dilute at the time the trypsin solution to 1 ng/μL. Add 1 μL of trypsin solution at 1 μg/μL to 999 μL of 50 mmol/L ammonium bicarbonate.
4. Dilute pure formic acid (HPLC-MS grade) 1 to 10 with water (HPLC-MS grade) to obtain a 10% formic acid solution.
5. Seal the sample plate with Agilent sealing foil in polyethylene terephthalate.

2.3.6 LC-MS Analysis

All the chemicals mentioned in this section must be HPLC-MS grade.

1. QC HSA: dilute digested HSA to 1 fmol/μL with BSA dilution solution.
2. Digest of BSA (Bovine Serum Albumin) is provided by Agilent. Resuspend the dry digest at 1 pmol/μL with 500 μL of 15% acetonitrile/85% water containing 0.1% formic acid. Dilute this solution 1/100 with the same buffer (15% acetonitrile/85% water containing 0.1% formic acid) to obtain 10 fmol/μL of "BSA dilution solution".
3. Digest of HSA (Human Serum Albumin) is provided by Agilent. Resuspend dry peptides with 50 μL of BSA dilution solution to obtain HSA at a concentration of 10 pmol/μL. Make a serial dilution of HSA solution with "BSA dilution solution" to obtain a final a final sample of 1 fmol/μL of HSA dilute in 1 fmol/μL of BSA digest.
4. Phase A: ultrapure water with 0.1% formic acid. Prepare 1 L by adding 1 mL of pure formic acid and 999 mL of water.
5. Phase B: methanol with 0.1% formic acid. Prepare 1 L by adding 1 mL of pure formic acid and 999 mL of methanol.
6. The liquid chromatography system used consists of a column (Zorbax 300 SB-C18 3.5 μm 150 mm × 1.0 mm) directly connect to the ion source.

2.3.7 Data Analysis

Skyline software v3.3 (Skyline Targeted Mass Spec Environment, MacCoss Lab, Seattle, WA, USA).

2.4 Other Supplies

1. CSF samples correspond to patients who had undergone a lumbar puncture to investigate memory complaints in the neurology department of the CHU of Montpellier. CSF samples were address to the laboratory to measure the concentration of

CSF biomarkers: amyloid peptides and tau proteins. The samples were collected following approved guideline [9]).

2. 50 mL polypropylene tubes (ref CLS430290-500EA, Corning).
3. 10 mL polypropylene tubes (ref 62 610 201, Sarstedt) (**Note 3**).
4. 0.5 and 1.5 mL low-adsorption polypropylene tubes (LoBind, colorless, Eppendorf).

2.5 Other Equipment

1. A fume hood.
2. A refrigerated centrifuge for 1.5 mL tubes.
3. A refrigerated centrifuge for 10 mL tubes.
4. A vibrating platform shaker—Titramax 1000 (Heidolph Instruments, Schwabach, Germany).
5. A vortex mixer.
6. Acid-resistant CentriVap Vacuum concentrators.
7. An analytical balance model Sartorius CPA224S-OCE.
8. An Eppendorf 5804 centrifuge for plates.
9. An extraction plate manifold connected to a vacuum pump.
10. Pipet-aid XP Drummond.
11. Multichannel pipette 20–200 μL (ref 613-5252, VWR).
12. Water bath 37 °C.

3 Method

3.1 Samples

1. Lumbar puncture (LP) is performed under standardized conditions at the end of the morning. LP is carried out at the L3/L4 or L4/L5 inter-space after ruling out the occurrence of any potential contraindications.
2. Collect CSF samples directly in 10 mL polypropylene tubes to prevent variations in the adsorption of biomarkers to the container surface (**Note 3**).
3. Transfer the sample tubes at 4 °C/on ice to the biochemical laboratory within 4 h.
4. Centrifuge the CSF samples at 1000 ×*g* for 10 min at 4 °C.
5. Aliquot the supernatant by 0.4 mL in 0.5 mL LoBind tubes of in 0.5 mL–1 mL in 1.5 mL LoBind tube.
6. Store at -80 °C before use.

3.2 Preparation of Immunoassay Internal Quality Controls (iQC)

1. Select several frozen CSF samples with known tau values close to the clinical cutoff (**Note 4**).
2. After thawing, pool the CSF in a single 50 mL tube and homogenize by inversion several times.
3. Centrifuge at 1500 ×*g*, 10 min at 4 °C.
4. Aliquot by 0.4 mL in 0.5 mL LoBind tubes of in 0.5 mL–1 mL in 1.5 mL LoBind tube.
5. Store at -80 °C before use.

3.3 Immunoassays (ELISA, CLEIA)

1. Follow the daily/monthly maintenance of the analyzers as recommended by the providers.
2. For Lumipulse, run a tau/ptau calibration using calibrators set and protocols before the first assay, when changing lot numbers, if the previous calibration has been performed more than 30 days before, if QC do not pass (see below), after service maintenance.
3. Select the frozen clinical CSF samples to be measured.
4. Select the right protocol on each analyzer (tTau or pTau) and program the singlicate or duplicate measurement of the samples and the iQC.
5. Plan the measurement of the QCs provided with the kits before and after the series of samples.
6. Along with an iQC aliquot, thaw the samples at room temperature (20–25 °C) for 30 min.
7. Vortex at least 10 s at maximum speed the samples before placing the Eppendorf on the analyzers using adaptors. A volume of 0.3 mL is necessary for each duplicate measurement.
8. Perform the assay and keep in mind the following general considerations:
   - Avoid bubbles in samples and reagents.
   - Unless indicated, bring reagents (calibrators…) at room temperature before use.
   - Avoid freeze-thaw cycles of samples, QC, calibrators, or reagents.
   - Monitor the room temperature as it may impact the analyses.
   - Handle any waste in accordance with your local regulations.
   - When any liquid is spilled, wipe and disinfect the area with appropriate solutions (disinfectant…) in accordance with your local regulations.
   - Avoid any dilution of the samples that is not recommended by the kit provider (linearity is not warranty in relation with matrix effect).

- Even if using the kits as recommended by the manufacturer, it is wise to verify its reference range by comparing in the studied population at least 20 samples in each clinical group (disease and control).

3.4 Quality Control (QC)

- The first assay validation criteria is represented by the values of the kit QCs (2 levels for ELISA/Euroimmun and 3 levels for CLEIA/Lumipulse) that need to be in the expected range.
- For the ELISA/Euroimmun system, the calibration curve has to follow different validation criteria: CV $< 20\%$; increasing optical density (OD) of standards 1 to 6 (S1–S6); minimum OD values for S1 and S6.
- To control the value of the iQC use the Westgard Rules [10] as follows: compute the mean and standard deviation (SD) of the 5 or the 10 previous iCQ values obtained. If the new value is comprised between the previous mean $\pm 2 \times$ SD the criteria is validated. If the new value is not comprised between the previous mean $\pm 3 \times$ SD, the criteria is not validated and the series values cannot be used. If the new value is intermediate between the previous mean $\pm 2 \times$ SD and $\pm 3 \times$ SD, the values might be use if this is not the same time in a row that this situation is present (Fig. 5.1, **Note 5**).

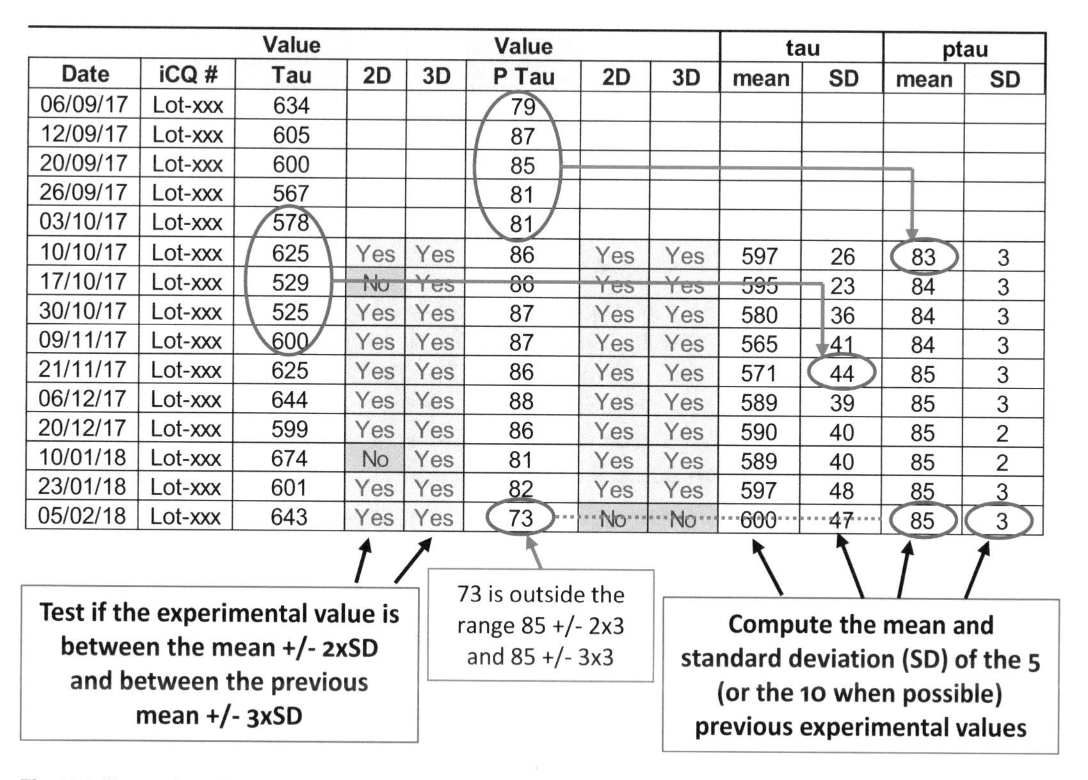

		Value			Value			tau		ptau	
Date	iCQ #	Tau	2D	3D	P Tau	2D	3D	mean	SD	mean	SD
06/09/17	Lot-xxx	634			79						
12/09/17	Lot-xxx	605			87						
20/09/17	Lot-xxx	600			85						
26/09/17	Lot-xxx	567			81						
03/10/17	Lot-xxx	578			81						
10/10/17	Lot-xxx	625	Yes	Yes	86	Yes	Yes	597	26	83	3
17/10/17	Lot-xxx	529	No	Yes	86	Yes	Yes	595	23	84	3
30/10/17	Lot-xxx	525	Yes	Yes	87	Yes	Yes	580	36	84	3
09/11/17	Lot-xxx	600	Yes	Yes	87	Yes	Yes	565	41	84	3
21/11/17	Lot-xxx	625	Yes	Yes	86	Yes	Yes	571	44	85	3
06/12/17	Lot-xxx	644	Yes	Yes	88	Yes	Yes	589	39	85	3
20/12/17	Lot-xxx	599	Yes	Yes	86	Yes	Yes	590	40	85	2
10/01/18	Lot-xxx	674	No	Yes	81	Yes	Yes	589	40	85	2
23/01/18	Lot-xxx	601	Yes	Yes	82	Yes	Yes	597	48	85	3
05/02/18	Lot-xxx	643	Yes	Yes	73	No	No	600	47	85	3

Test if the experimental value is between the mean +/- 2xSD and between the previous mean +/- 3xSD
73 is outside the range 85 +/- 2x3 and 85 +/- 3x3
Compute the mean and standard deviation (SD) of the 5 (or the 10 when possible) previous experimental values

Fig. 5.1 Illustration of an excel sheet used for the visual validation of the Westgard criteria

Table 5.1
Gravimetric preparation of Tau calibration standards

0.5% goat serum	N14 tau at 100 ng/mL	C14N expected mean (ng/mL)
1.915 g	4.085 g	0.32
1.452 g	4.548 g	0.47
1.169 g	4.831 g	0.62
1.911 g	4.089 g	0.77
3.848 g	2.152 g	1.13
3.549 g	2.451 g	3.15
2.093 g	3.907 g	7.71
1.417 g	4.583 g	11.84
3.090 g	2.910 g	15.5
4.082 g	1.918 g	31.96

3.5 Mass Spectrometry

3.5.1 Tau Standards Preparation

1. Prepare gravimetrically ten calibration standards by adding a fixed amount of 15N tau to variable amounts of 14N tau (Table 5.1).
2. Dilute the standards 1/100 with 50 mmol/L ammonium bicarbonate and 1 mmol/L BSA to obtain the "Working solution 1". Dilute 1 g of "Stock solution" with 99 g of the mix "50 mmol/L ammonium bicarbonate; 1 mmol/L BSA".
3. Dilute secondly with 0.5% goat serum to obtain the "Working solution 2". Dilute 1 g of the "Working solution 2" with 99 g of 0.5% goat serum.
4. 14N tau ranged from 0.32 to 31.96 ng/mL. Calibration points are 0.32; 0.47; 0.62; 0.77; 1.13; 3.15; 7.71; 11.84; 15.5; and 31.96 ng/mL. Table 5.1 describes the dilution protocol. Briefly, a serial dilution by mixing "working solution 2" and "0.5% goat serum" in different proportion give each calibration point. Analyze each point in triplicate.
5. In these calibration points, reach a final concentration of 4 ng/mL for 15N tau.

3.5.2 Protein Precipitation

1. Unfreeze 500 μL of CSF sample on ice.
2. Spike a final concentration of 4 ng/mL for 15N tau by adding 20 μL of the "working solution 2".
3. Vortex on ice for 5 min at 1000 rpm.

4. Optionally, generate the calibration curve in parallel. Unfreeze 500 μL of each calibrants generated before (Subheading 3.2.4).
5. Acidify the sample by adding 25 μL of 70% perchloric acid.
6. Vortex for 10 s at maximum speed.
7. Wait 15 min on ice (**Note 6**).
8. Centrifuge at 1600 ×*g* for 15 min at 4 °C.
9. Carefully collect the supernatant and transfer on a new 1.5 mL polypropylene tubes (**Note 7**).

3.5.3 Solid Phase Extraction

This step must be performed under a fume hood because of the toxic organic and acidic vapor release. SPE steps are performed using an extraction plate manifold connected to a vacuum pump. To aspirate the liquid, the depression is gradually increased to 55 kPa (**Note 8**). The liquid is collected below the SPE plate in the collection plate, except during the elution step.

1. Wash the phase with 300 μL of solvent A. Aspirate.
2. Condition the phase with 500 μL of solvent B. Aspirate.
3. Load the sample by transferring all the sample volume into the well. Wait 2 min before and aspirate gently for 30 s until no liquid passes through (**Note 9**).
4. Wash the sample with 500 μL of solvent B. Aspirate.
5. Repeat **step 4**.
6. Change the collection plate by the elution plate below the SPE plate.
7. Elute the cleaned sample with 100 μL of solvent C. Wait 2 min before aspirating it gently for 30 s until all the liquid passes through (**Note 9**).

3.5.4 Sample Drying

1. Switch on the concentrator system.
2. Put the Elution plate in the centrifugal unit without covering it. Equilibrate with an equivalent plate.
3. Select the centrifugal program corresponding to 30 °C for 1 h (**Note 10**).
4. Switch on the centrifugal unit and open the depression valve.

3.5.5 Protein Digestion

1. Resuspend the sample directly with 40 μL of the trypsin solution at 1 ng/μL.
2. Digest the sample at 37 °C with stirring at 400 rpm during 24 h.
3. Stop the digestion by acidification with 5 μL of 10% formic acid.

4. Seal the sample plate for storage and also to avoid evaporation on LC autosampler.

3.5.6 LC-MS Analysis

1. Remove gas from solvents A and B. Purge the pump. Run two blank runs to equilibrate all the fluidic parts of the LC system.
2. LC-MS performances are checked with the QC HSA which is injected prior to sample analysis.
3. Store the sample in the autosampler with control temperature at 4 °C.
4. Use following parameters to the LC separation: flow rate 50 μL/min, oven temperature 60 °C.
5. Inject 20 μL of the sample into the chromatographic system in a 21-min LC run (Table 5.2).
6. Mass spectrometer operates in MRM scan mode (Table 5.3).

3.5.7 Data Analysis

Table 5.2
LC and autosampler parameters

Autosampler	Tray temperature	4 °C	
Injection mode	Volume	20 μL	
	Needle wash	10s on flush port	
	Draw speed	100.0 μL/min	
	Eject speed	100.0 μL/min	
	Draw position	0.0 mm	
	Equilibration time	2.0 s	
	Sample flush out factor	5 times injection volume	
	Well bottom sensing	Yes	
Chromatographic system	Column	Zorbax 300 SB-C18 3.5 μm 150 mm × 1.0 mm	
	Column temperature	60 °C	
	Flow rate	50 μL/min	
	Phase A	Ultrapure water with 0.1% formic acid	
	Phase B	Methanol with 0.1% formic acid	
	Gradient	Time (min)	Phase B (%)
		0	2
		2	2
		2.1	10
		15.1	70
		15.2	90
		16.2	90
		16.3	2
		31.3	2

Table 5.3
ESI source and MS parameters

ESI source	Ion Source	JetStream			
	Capillary tension	2500 V			
	Gas flow	16 L/min			
	Gas temperature	140 °C			
	Sheath gas flow	7 L/min			
	Sheath gas temperature	250 °C			
	Nebulizer	40 psi			
MS parameters	Mode	Positive			
	Ion funnel HP RF	150 V			
	Ion funnel LP RF	110 V			
	Cell accelerator voltage	5 V			
	MS1 resolution	Wide			
	MS2 resolution	Wide			
MRM mode	**Compound name**	**Precursor ion**	**Product ion**	**Product ion type**	**Collision energy**
	14N tau: GAAPPGQK	363.4	597.7*	y_6^+	4
			526.6	y_5^+	4
			429.5	y_4^+	12
	15N tau: GAAPPGQK	368.4	605.6*	y_6^+	4
			533.6	y_5^+	4
			435.5	y_4^+	12

* are the quantifier ions

Table 5.4
Skyline settings

Peptide Settings	Digestion	Enzyme	Trypsin [KR|P]
		Maximum missed cleavages	1
		Background proteome	None
	Prediction	Retention time predictor	None
		Ion mobility predictor	None
		Use measured retention times when present	None
		Use spectral library ion mobility values when present	None
	Filter	Minimum length	6 AA
		Maximum length	25 AA
		Exclude N-terminal AA	1
		Exclude potential ragged ends	None
		Exclude peptides containing specific sequence	None
		Auto-select all matching peptides	None
	Library	None	
	Modifications	Structural modifications	None
		Isotope label type	heavy
		Isotope modifications	Label:15N
		Internal standard type	Heavy
	Quantification	Regression Fit	Linear
		Normalization Method	Ratio to heavy
		Regression Weighting	1/x
		MS level	All
		Units	ng/mL
Transition settings	Prediction	Precursor mass	Average
		Product ion mass	Average
		Collision energy	Agilent QQQ
		Declustering potential	None
		Optimization library	None
		Compensation voltage	None
		Use optimization values when present	Yes
		Optimize by	Transition
	Filter (peptides)	Precursor charges	1, 2, 3

	Ion charges	1, 2, 3
	Ion types	y, b
	Product ion selection	From ion 3
		To last ion -1
		Special ions: N-terminal to pro
	Precursor m/z exclusion window	4 m/z
	Auto-select all matching transitions	Yes
Library	Ion match tolerance	0.05 m/z
	If a library spectrum is available, pick its most intense ions	Yes
	Pick	10 product ions
		From filtered ion charges and types
Instrument	Minimum m/z	300
	Maximum m/z	1050
	Dynamic min product m/z	None
	Method match tolerance m/z	0.01 m/z
Full-scan		None

1. Skyline require to fix settings at the beginning of the project to correctly process the raw data.
2. Peptide and Transition Settings are listed on Table 5.4.
3. Load Agilent raw data directly on Skyline. Area ratio could be directly compared between samples or against the calibration curve.

4 Notes

1. Store pure TFA bottles at 4 °C to prevent hydrolysis. The bottle must be opened carefully under a fume hood to prevent skin/eye irritation with acidic vapor. An intermediate dilution of 10% TFA is performed in order to reduce TFA toxicity. This solution is stable when stored at 4 °C.
2. Lyophilized trypsin powder is electrostatic. Carefully open the trypsin vial. Store the trypsin at -20 °C at this concentration to preserve enzymatic activity for months. Aliquot it to avoid freeze/thaw cycles.
3. Since CSF samples contain much less protein than serum or plasma samples, the absorption of proteins to tube walls represents an important source of bias. To avoid loss of biomarkers as reported to occur [11] use specific polypropylene tubes.
4. In addition to an iCQI which expected value is close to the clinical cutoff and additional iCQ in the lower or the higher rage is also recommended.
5. Prepare enough iCQ aliquots (>50) knowing that when you change of iCQ lot you will have to run the new at least lot 5 times to validate its value. The Fig. 5.1 corresponds to an excel sheet that illustrates the Westgard Rules.
6. A white film must appear inside the liquid. Otherwise, an extra waiting time of 15 min can be added. If the white film does not develop, something has gone wrong with either the sample or the acidification step.
7. The supernatant has to be removed very carefully. The tube must be removed from the centrifuge slowly to avoid mixing. Aspirate slowly with the pipette without aspiring any particles. If this is not possible, the sample can be re-centrifuged.
8. Do not over-dry the well by aspiring for too long or too strongly. This is a crucial point for obtaining good and reproducible chromatographic results.

9. These precautions are necessary to ensure a long enough contact time between the molecules (in the sample) and the chromatographic phase.
10. A low-temperature (30 °C) drying step prevents protein modifications due to the loss of water or deamidation, for example. The drying step usually takes 1 h, but the time can be increased if necessary.

References

1. Drubin DG, Kirschner MW (1986) Tau protein function in living cells. J Cell Biol 103(6 Pt 2):2739–2746. Epub 1986/12/01
2. Wang JZ, Xia YY, Grundke-Iqbal I, Iqbal K (2012) Abnormal hyperphosphorylation of tau: sites, regulation, and molecular mechanism of neurofibrillary degeneration. J Alzheimers Dis 33:S123–S139. Epub 2012/06/20
3. Iqbal K, Flory M, Khatoon S, Soininen H, Pirttila T, Lehtovirta M et al (2005) Subgroups of Alzheimer's disease based on cerebrospinal fluid molecular markers. Ann Neurol 58 (5):748–757. Epub 2005/10/26
4. Vanmechelen E, Vanderstichele H, Davidsson P, Van Kerschaver E, Van Der Perre B, Sjogren M et al (2000) Quantification of tau phosphorylated at threonine 181 in human cerebrospinal fluid: a sandwich ELISA with a synthetic phosphopeptide for standardization. Neurosci Lett 285(1):49–52. Epub 2000/05/02
5. McKhann GM, Knopman DS, Chertkow H, Hyman BT, Jack CR Jr, Kawas CH et al (2011) The diagnosis of dementia due to Alzheimer's disease: recommendations from the National Institute on Aging-Alzheimer's Association workgroups on diagnostic guidelines for Alzheimer's disease. Alzheimers Demen 7 (3):263–269. Epub 2011/04/26
6. Mattsson N, Zetterberg H, Hansson O, Andreasen N, Parnetti L, Jonsson M et al (2009) CSF biomarkers and incipient Alzheimer disease in patients with mild cognitive impairment. JAMA 302(4):385–393. Epub 2009/07/23
7. Gabelle A, Dumurgier J, Vercruysse O, Paquet C, Bombois S, Laplanche JL et al (2013) Impact of the 2008–2012 French Alzheimer plan on the use of cerebrospinal fluid biomarkers in research memory center: the PLM Study. J Alzheimers Dis 34(1):297–305. Epub 2012/11/29
8. Bros P, Vialaret J, Barthelemy N, Delatour V, Gabelle A, Lehmann S et al (2015) Antibody-free quantification of seven tau peptides in human CSF using targeted mass spectrometry. Front Neurosci 9:302. Epub 2015/09/22
9. Del Campo M, Mollenhauer B, Bertolotto A, Engelborghs S, Hampel H, Simonsen AH et al (2012) Recommendations to standardize preanalytical confounding factors in Alzheimer's and Parkinson's disease cerebrospinal fluid biomarkers: an update. Biomark Med 6 (4):419–430. Epub 2012/08/25
10. Westgard JO. Westgard rules. https://www.westgard.com/mltirule.htm
11. Perret-Liaudet A, Pelpel M, Tholance Y, Dumont B, Vanderstichele H, Zorzi W et al (2012) Cerebrospinal fluid collection tubes: a critical issue for Alzheimer disease diagnosis. Clin Chem 58(4):787–789. Epub 2012/02/11

Chapter 6

CSF RT-QuIC and the Diagnosis of Creutzfeldt–Jakob Disease

Alison J. E. Green and Neil I. McKenzie

Abstract

Real-time Quaking-Induced Conversion (RT-QuIC) is a protein aggregation assay that exploits the ability of mis-folded proteins, such as prion protein, to induce self-aggregation. By introducing heating and shaking steps the aggregated proteins can be fragmented to induce further aggregation and the resultant aggregated protein forms fibrils that can be detected using a fluorescent dye that binds to amyloid fibrils. This results in an assay that can amplify small amounts of mis-folded proteins in biological fluids such as cerebrospinal fluid (CSF) and by using a fluorescent spectrophotometer monitor the formation of aggregated fibrils in real time. The development of CSF RT-QuIC has been a major advance in the pre-mortem diagnosis of sporadic Creutzfeldt–Jakob disease (sCJD). It has a sensitivity of 90–94% and specificity approaching 100%. From January 2017 a positive CSF RT-QuIC has been included into the diagnostic criteria of sCJD.

Key words CSF, RT-QuIC, Sporadic CJD, Prion disease, Prion protein, Prion, CJD biomarkers

Abbreviations

Abs_{280}	Absorbance at 280 nm wavelength
BH	Brain Homogenate
CJD	Creutzfeldt–Jakob disease
CSF	Cerebrospinal fluid
E. coli	*Escherichia coli*
EDTA	Ethylenediaminetetraacetic acid
FPLC	Fast Protein Liquid Chromatography
Ha	Hamster derived
Ham FL	Hamster full length (residues 23–231)
Ham-Sh	Hamster-Sheep chimeric protein
Hu	Human derived
Hum FL	Human full length
IB	Inclusion Bodies
IMAC	Immobilized Metal Affinity Chromatography
LB	Lysogeny Broth
MBSC	Microbiological Safety Cabinet

Charlotte E. Teunissen and Henrik Zetterberg (eds.), *Cerebrospinal Fluid Biomarkers*, Neuromethods, vol. 168,
https://doi.org/10.1007/978-1-0716-1319-1_6,

MM1	Genotype that is homozygous for methionine at codon 129, Type 1
MWCO	Molecular-weight cut-off
NTA	Nitrilo-triacetic acid
OETB	Overnight Express Instant TB Medium
PBS	Phosphate-Buffered Saline
PK	Proteinase K, a broad spectrum serine protease
PrP	Prion Protein
PrPres	PK-resistant Prion Protein
PrP^{Sc}	Prion Protein (insoluble/disease causing/"scrapie" form)
RCF	Relative Centrifugal Force
recPrP	Recombinant Prion Protein
rfu	Relative Fluorescence Units, a measure of fluorescence
RT-QuIC	Real-Time Quaking-Induced Conversion (assay)
sCJD	Sporadic Creutzfeldt–Jakob disease
SDS	Sodium Dodecyl Sulfate
TB	Terrific Broth
ThT	Thioflavin T., a solvatochromic fluorescent dye
vCJD	Variant form of Creutzfeldt–Jakob disease
w/v	Weight to volume, e.g., *X g per Y mL*
w/w	Weight to weight, e.g., *X mL per Y mL*

1 Introduction

Prion diseases are a family of fatal neurodegenerative diseases that affect both animals and humans, which are associated with the deposition within the central nervous system of a mis-folded form (PrP^{Sc}) of a normal constitutive protein called prion protein (PrP^{C}). The most common form of prion disease affecting humans is sporadic Creutzfeldt–Jakob disease (sCJD) which is associated with a rapidly progressing cognitive decline with death occurring within 6 months [1]. Neuropathologically sCJD is characterized by severe neuronal loss with spongiform change, marked astrocytosis, and the presence of PrP^{Sc}.

PrP^{Sc} is able to bind to the normal form of prion protein PrP^{C} and induce it to change shape, in a self-propagating manner. The PrP^{Sc} thus produced can then induce further conformational change resulting in an accumulation of PrP^{Sc} that then aggregates to form amyloid deposits within the central nervous system.

Until recently, the pre-mortem diagnosis of sCJD depended on the clinical features and the results of investigations such as MRI, EEG, and the presence of 14-3-3 in the cerebrospinal fluid (CSF) [2, 3]. 14-3-3 is a neuronal protein that is released into the CSF following acute neuronal damage, and as such is not specific for sCJD. In addition the MRI or EEG changes associated with sCJD are not specific for this disorder.

To improve the specificity of the diagnostic tests for sCJD, methodologies which exploited the ability of the disease-associated PrP^{Sc} to induce aggregation of normal cellular PrP^{C} have been developed. The first of these so-called protein aggregation assays to be widely used was protein mis-folding cyclic amplification (PMCA) which used brain homogenates as a source of PrP^{C} and Western blotting as a detection system for the abnormal PrP^{Sc} produced [4]. PMCA was able to detect very small amounts of PrP^{Sc} present in blood; however, the need for a second detection system limited its usefulness in a clinical setting.

Real-time Quaking-Induced Conversion (RT-QuIC) has two major advantages over PMCA; firstly, it combines both amplification and detection in one process and secondly it utilizes a 96-well microtiter plate format which facilitates its use as a large throughput assay [5–7]. In brief, CSF samples are added to a detergentless buffer that contains recombinant PrP as a substrate and a fluorescent marker, Thioflavin T (ThT). The abnormal PrP^{Sc} (seed) in the sCJD CSF sample binds to the recombinant PrP (rec PrP) (substrate) and induces it to change shape and aggregate. This aggregation is promoted by heating the samples and by intermittent shaking. The aggregated PrP^{Sc} binds ThT and causes a change in the emission spectra which is monitored continuously by fluorescence spectroscopy. The amount of aggregation can therefore be followed by exciting the ThT dye at 450 nm and recording the fluorescence emission at 480 nm.

There is an initial lag phase of approximately 30 h during which small amounts of dimer, trimer, and small aggregates are formed, which is followed by a rapid increase in fluorescence once these small aggregates coalesce to form fibrils that rapidly increased in size. Maximal fluorescence is achieved by 90 h. The resultant sigmoid curve is consistent with a nucleation-dependent polymerization model [8] (Fig. 6.1).

A number of different rPrP substrates have been used in RT-QuIC and the advantages of each have been described elsewhere [9]. A number of retrospective and prospective studies investigating the sensitivity and specificity of CSF RT-QuIC for the diagnosis of sCJD have been reported [6, 10–17]. The sensitivity of CSF RT-QuIC for the diagnosis of sCJD is between 90–95% and the specificity is approaching 100% [6, 10–17]. Two large international ring-trials have shown that there is good concordance of results between laboratories in different countries [15, 18]. This robustness and high degree of sensitivity and specificity has led to a positive CSF RT-QuIC result being included in the diagnostic criteria for the clinical diagnosis of sCJD (www.cjd.ed.ac.uk).

Sensitivity of RT-QuIC in genetic CJD depends on the particular mutation [19]. CSF samples from patients with variant CJD (vCJD) are not positive in the RT-QuIC system described above. Negative results are obtained for “variable protease-sensitive prionopathy” (VPSPr) under these conditions.

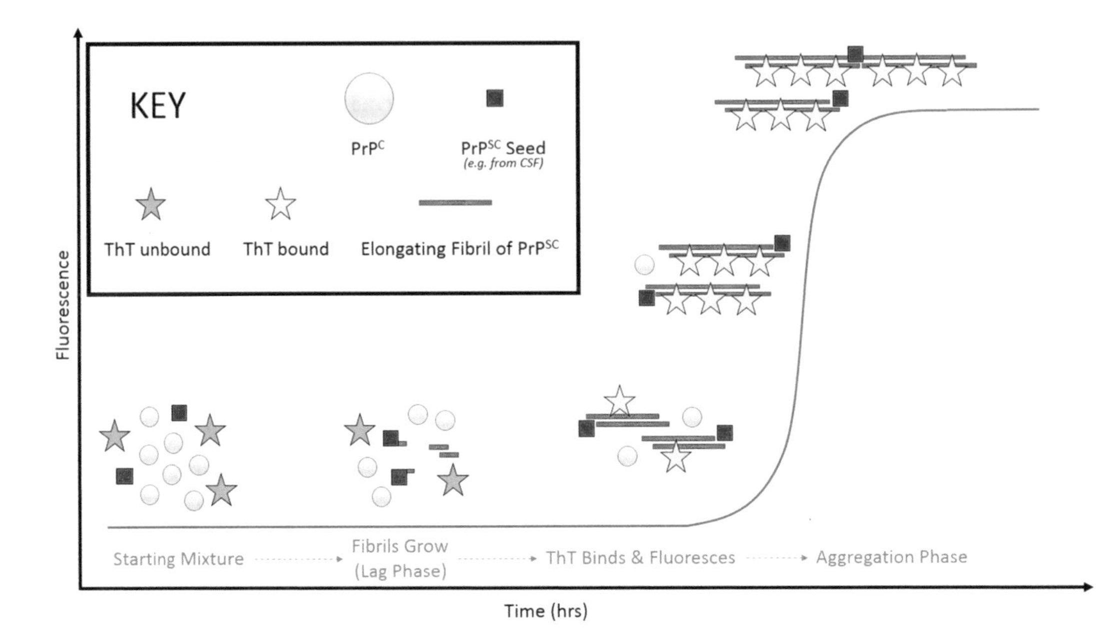

Fig. 6.1 RT-QuIC seeded with sCJD brain homogenate. Misfolded PrP^{SC} seeds template the aggregation of PrP^{C} into PrP^{SC} fibrils, which have a large beta-sheet component. Thioflavin T (ThT) binds to beta-sheet regions of protein and can be induced to fluoresce using laser light at an excitation wavelength maximum of $\lambda = 450$ nL. The rate of fibril formation can thus be followed by measuring the (bound) ThT emission wavelength maximum at $\lambda = 480$ nm

The information described below is based on published methods [16, 20] along with the surveillance data and experience of the National CJD Research and Surveillance Unit (NCJDRSU), Edinburgh, UK.

1.1 A Brief RT-QuIC Work Plan

Each step is explained more fully in the subsequent sections:

1. Thaw seeds (CSFs and Brain Homogenates) and PrP at room temperature.
2. Calculate each master mix "recipe."
   i.e., *How many control samples/30μL CSF samples/15μL CSF samples will there be?*
3. Prepare 10 mL of each component buffer of the master mix and filter (0.2μ).
4. Partially prepare each master mix—do not add the ThT, water or PrP^{C} yet.
5. Spin filter the freshly thawed PrP^{C} through a 100 kDa spin filter to remove any aggregates.
6. Determine the filtered PrP^{C} concentration by measuring Abs 280 nm of the solution.
7. Calculate how much PrP^{C} and water should be added to each master mix.
8. Add the ThT and water to the master mixes.

9. Vortex the mixtures, then add the PrP^C with gentle pipetting.
10. Add the appropriate volume of master mix to the appropriate wells, avoiding bubbles.
11. Add the appropriate volume of seed to the appropriate wells, avoiding bubbles.
12. Seal the plate with adhesive plastic film.
13. Load the plate into the plate reader and run a programmed script which specifies how the plate reader should shake and read the plate.

2 Materials

1. Consumables for the RT-QuIC assay (Fig. 6.2)
2. Apparatus for RT-QuIC (Fig. 6.3)
3. Overview of RT-QuIC controls

   In the CSF RT-QuIC assay, the control samples used are brain homogenates* and the test samples are CSF. Brain homogenates should be obtained from a tissue bank and have been collected with ethical approval.

Reagent	Concentration	Volume per well
Hamster Prion Protein	0.3-0.5 mg/ml	20 – 30 µl (10 µg after spin filtration)
Phosphate Buffered Saline (PBS), 5x conc	50 mM Phosphate 770 mM NaCl	20 µl
0.1% SDS in a 1 x PBS solution ‡	10 mM Phosphate 154 mM NaCl 0.1% SDS	
Sodium Chloride Solution	2 M	8.5 µl
EDTA	100 mM	1 µl
ThT *(make 10 mM and dilute 10x)*	1 mM	1 µl
Water, molecular biology grade	-	(100 - 30.5 - recPrP – Sample) µl
Samples	Concentration	Volume per well
Sudden-Death Brain Homogenate	Diluted as per CJD brain homogenate	2 µl
Alzheimer's Diseased Brain Homogenate	Diluted as per CJD brain homogenate	2 µl
sCJD Brain Homogenate	50 fg Hu PrPres/µl†	2 µl
CSF sample	Neat	30 or 15 µl

Fig. 6.2 Overview of materials required for RT-QuIC. †Concentrations are determined using Western Blot. Quantitation is calculated as detailed in the section titled "Preparing the Brain Homogenate Controls". ‡Not used in the RT-QuIC assay, only used to determine recPrP concentration

Apparatus	Amount
Alkali resistant carboy (for disposal of waste)	As required
BMG Labtech FLUOstar Omega or Optima, including computer and software	1
Clear polyolefin adhesive tape for sealing 96 well plate	1
Eppendorfs, 2 ml	As required
Falcon Tubes, 15 and 50 ml	As required
Level 2 MBSC inside a Category 3 laboratory*	1
Optical flat bottom plate, 96 well	1
Pipettes and sterile filter tips, ranging from 0.1 µl to 1 ml volume	As required
Sharpsafe (for disposal of pipette tips)	1
Spin filtration units, with 100 kDa MWCO	As required
Sterile filters, 0.22 µ porosity, with built-in Luer-lock screw attachment	As required
Syringes, polypropylene Luer-lock tips, capacity 10 ml	As required

Fig. 6.3 Overview of equipment required for RT-QuIC

Each control and CSF sample is run in quadruplicate (i.e., 4 wells allocated on the plate per sample) to reduce the risk of false positives biasing the test outcome.

These controls are as follows:

(a) Unseeded master mix

(b) 100 fg of PrP from sudden-death brain

(c) 100 fg of PrP from Alzheimer's diseased brain

(d) 100 fg of sCJD PrPres from MM1 genotype brain*

For the control samples, 2µL of brain homogenate is used to seed each quadruplicate well set. Out of the four control types, only the controls wells which were spiked with sCJD brain homogenate exhibit a large increase in rfu during the RT-QuIC assay. The rfu of the other control wells remain low and constant throughout the whole assay.

> * Requires a category 3 laboratory. An alternative positive control, suitable for use in a category 2 laboratory, is a diluted solution of CSF RT-QuIC reaction products.

(a) **Consumables for preparing RT-QuIC controls** (Fig. 6.4)

(b) **Soto conversion buffer** (Fig. 6.5)

(c) **Equipment for preparing RT-QuIC controls** (Fig. 6.6)

Material	Amount
Soto conversion buffer	10 ml
Alzheimer's Diseased Brain - frontal cortex	100 mg
Anti-prion protein antibody 3F4 epitope	100 µg
N2 supplement, 100x concentration	5 ml
Proteinase K	5 mg
Recombinant Human PrP	100 ng
Sporadic CJD Brain - frontal cortex	100 mg
Sudden-Death Brain - frontal cortex	100 mg

Fig. 6.4 Overview of reagents required for preparing RT-QuIC brain homogenate controls

- 10x PBS 1ml
- 0.5M EDTA 20ml (Final conc. = 1mM)
- 5M NaCl 0.3ml (Final conc. = 150mM)
- 10% (v/v) Triton X-100 500ml (Final conc. = 0.5%)
- Complete Protease Inhibitor Cocktail, EDTA free (Roche 11836170001) *[i.e. 1 tablet]*

Make up to 10ml with H_2O.

Incubate end over end at 4ºC until PI Tablet dissolves and then put on ice until use.

Fig. 6.5 Overview of reagents required for preparing Soto conversion buffer

Apparatus	Amount
Alkali resistant carboy (for disposal of samples, Eppendorfs and well plates)	1
Eppendorfs, 2 ml	As required
Equipment, software and reagents for running Western Blots	1
Equipment and reagents for running and visualising SDS-PAGE	1
Heating block	1
Level 2 MBSC	1
Pipettes and sterile filter tips, ranging from 0.1 µl to 1 ml	As required
Scalpel	1
Sharpsafe	1
Sterile filters, 0.22 µ porosity, with built-in Luer-lock screw attachment	As required
Syringes, polypropylene Luer-lock tips, capacity 10 ml	As required

Fig. 6.6 Overview of equipment required for preparing brain homogenate controls

2.1 Safe Handling and Disposal of sCJD Brain Homogenates

Important!
All work involving the handling of sCJD brain homogenates should be carried out in a class 2 MBSC within a category 3 laboratory.
This specifically includes, but is not limited to, the sCJD seed dilutions, loading of well plates with sCJD seeds and the disposal of sCJD contaminated materials after the experiment is complete. Disposable pipette tips can be sealed inside a Sharpsafe kept within the MBSC.
Eppendorfs and well plates can be disposed of by soaking in 2 M NaOH solution at room temperature for a minimum of 1 h in an alkali impervious container with an airtight seal.

3 Methods

3.1 Preparing the Brain Homogenate Controls

(a) Prepare Soto conversion buffer.

(b) Collect frontal cortex samples (around 100 mg each) from sudden-death, Alzheimer's diseased and sCJD brains.

(c) Prepare 10% w/v solutions of these brain samples in Soto conversion buffer by dissolving brain (100 mg) in buffer (0.9 mL).

(d) Prepare a reference recPrP sample containing a known concentration of recombinant Hu PrP.

(e) Collect a 100μL aliquot of the sCJD sample solution and digest it with Proteinase K (PK, 2μL, 2 mg/mL) at 37 °C for 1 h. The other controls are not treated with PK.

(f) Prepare between 10 and 12 serial 1:10 dilutions of the PK-digested sCJD brain sample.

(g) Run a Western blot using these dilutions against a known concentration of Hu PrP. Visualize bands using anti-prion protein antibody 3F4 epitope, in accordance with the manufacturer's instructions.

(h) Compare the combined band densities of each digested sCJD dilution (3 bands) to that of the known Hu PrP concentration (1 band) and calculate stock concentration.

(i) Dilute the sCJD stock as required in Soto conversion buffer to give a final Hu PrP concentration of 100 fg/2μL.

(j) Dilute the SD and AD brain homogenates by the same factor as was calculated for the sCJD brain homogenate.

The brain homogenate controls should be aliquoted out and stored at −80 °C until required.

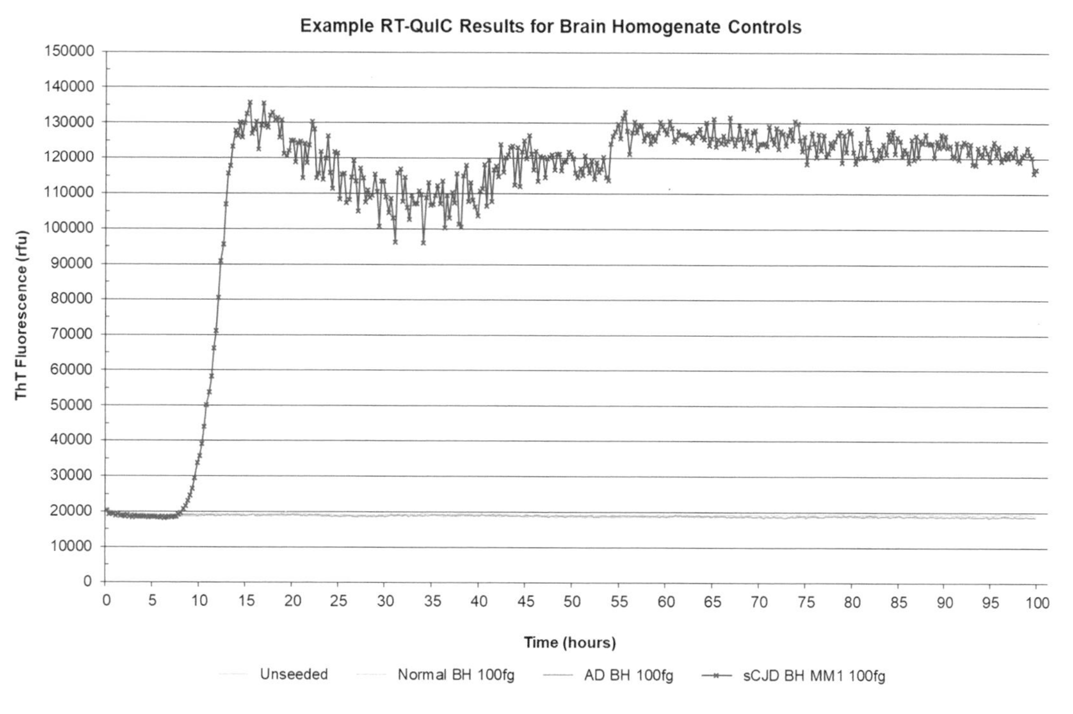

Fig. 6.7 Example traces illustrating how brain homogenate controls perform in RT-QuIC

3.2 Example Unseeded and Brain Homogenate Controls (Fig. 6.7)

3.2.1 Alternate Positive Control for Category 2 Laboratories

sCJD brain homogenate must only be handled in a secure category 3 laboratory. If only a category 2 laboratory is available, it is permissible to prepare a less infectious positive control as follows:

(a) Run a CSF RT-QuIC reaction as described in this protocol.

(b) Identify which wells gave a strong positive response, i.e., sCJD positive, >100,000 rfu signal.

(c) Inside an MSBC class II hood, remove the film seal (being careful of splashes) and pipette out the solutions in the positive wells.

(d) Dilute this reaction product ×1000 into 1× PBS, e.g., *if 4 wells have been selected, dilute a portion of the 400μL well product* (e.g., *5μL) into 1× PBS (5 mL).*

(e) Aliquot the diluted product into 50μL volumes.

(f) Freeze the aliquots and undiluted master stocks at −40 or −80 °C.

These can be freeze-thawed up to four times before a substantial deterioration in the positive control signal is observed.

3.3 Example of the Alternate Positive Control in RT-QuIC (Fig. 6.8)

3.3.1 Preparation of Master Mix

When preparing the master mix for a particular CSF volume (Fig. 6.9), the components of **ONE** well are (Fig. 6.9):

It is useful to prepare slightly more master mix than is required. For example, if measuring 16 wells, prepare enough master mix for 17 wells.

3.3.2 Preparation of Master Mix Components

One master mix is prepared for the brain homogenate controls and one master mix is prepared for each group of CSF volumes being tested.

i.e., *one master mix for all the samples to which brain homogenate seeds will be added, another master mix for all the wells where 30μL of CSF will be added and one master mix to which 15μL of CSF will be added.*

3.3.3 RT-QuIC Buffer Recipes

All buffers may be stored for up to 4 weeks once filtered.

Thioflavin T should be made up fresh for every plate and protected from light.

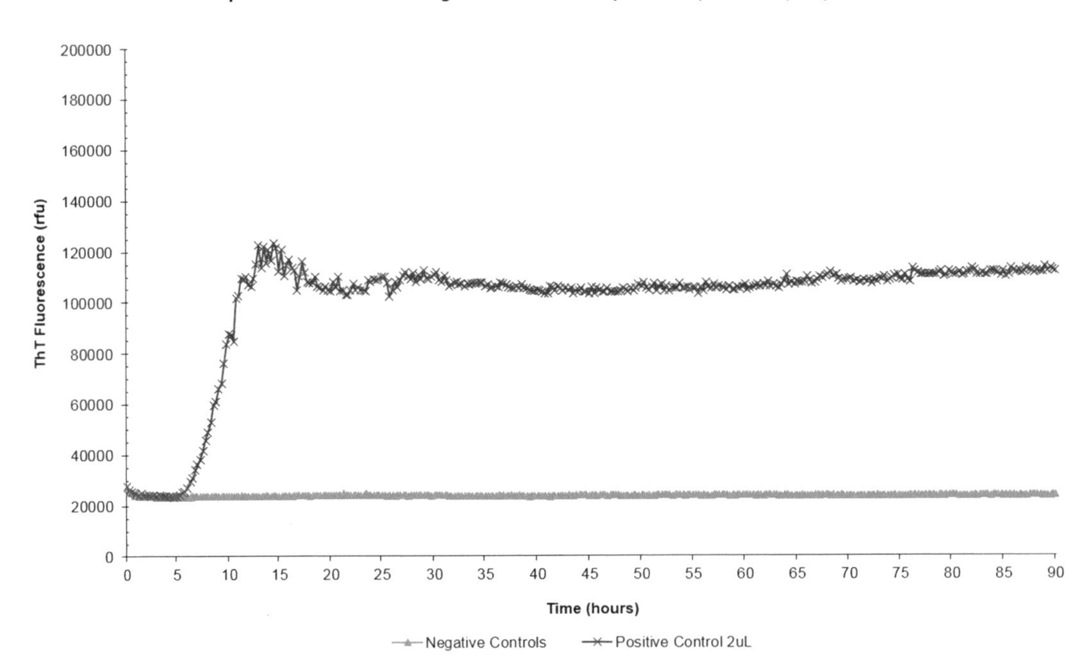

Fig. 6.8 Example trace illustrating how the alternate positive control performs in RT-QuIC

5 x PBS	=	20.0 μl
2M NaCl	=	8.5 μl
100mM EDTA	=	1.0 μl
1mM ThT	=	1.0 μl
Buffer components sum	=	30.5 μl

Seed volume = X μl

(i.e. 2 μl or 15 / 30 μl, depending on whether the master mix is for brain homogenate or CSF)

PrP^C = Y μl

(i.e. 10.0 μg after spin filtering through 100 kDa filter)

Water = (100 μl – Buffer components – Seed volume – PrP^C volume) μl

= (100 μl – Buffer components – Y – X) μl

Fig. 6.9 Overview of RT-QuIC components per well

5× PBS (10 mL)

Commercially available 10× PBS (5 mL) is diluted into demineralized water (4 mL).

Adjust final volume to ***0.01 L at pH*** *7.40.*

Final phosphate concentration = 50 mmol/L.

Final NaCl concentration = 770 mmol/L.

2 M NaCl (10 mL)

Commercially available 5 M NaCl solution (4 mL) is diluted into demineralized water (5 mL).

Adjust final volume to ***0.01 L at pH*** *7.40.*

Final NaCl concentration = 2 mol/L.

100 mM EDTA (10 mL)

Dissolve disodium Ethylenediaminetetraacetic acid, EDTA (37.2 mg, 100μmol, MW = 372.24 g/mol) in demineralized water (9 mL)

*Adjust final volume to **0.01 L at pH 8.00.***

1 mM ThT (10 mL)

Dissolve Thioflavin T (32.0 mg, 100μmol, MW = 318.86 g/mol) in demineralized water (9 mL) to afford a 10 mM solution.

Wrap the fluorescent yellow solution container in tinfoil to protect the contents from light.

On a benchtop shaker, shake the mixture at room temperature for ≥10 min.

Filter the solution through a 0.22μ filter and dilute x10 before use to afford a 1 mM ThT solution, e.g., *filtered 10 mM ThT solution (1 mL) diluted into demineralized water (9 mL).*

*Adjust final volume to **0.01 L***

3.3.4 Example Calculation

Here is an example calculation where a master mix is being prepared for 17 control wells. The same procedure can be followed to calculate master mix recipes for the CSF volumes being tested (Fig. 6.10).

1. The following volumes are combined (Fig. 6.10):
2. The recPrP is thawed and spin-filtered through a 100 kDa MWCO filter.
3. A sample of the freshly spin-filtered recPrP stock (100μL) is diluted by a factor of 10 into 1× PBS containing 0.1% SDS (0.9 mL) to allow the concentration of the neat stock to be determined. The SDS is used here to stop PrP^C sticking to the cuvette walls.

5 x PBS = 17 x 20 μl	= 340.0 μl
2M NaCl = 17 x 8.5 μl	= 144.5 μl
100mM EDTA = 17 x 1 μl	= 17.0 μl
1mM ThT = 17 x 1 μl	= 17.0 μl
	Sum = 518.5 μl

When calculating how much water to add in step 4, remember to account for the seed volume, i.e.

BH = 17 x 2 μl seed	= 34.0 μl
Sum = 518.5 μl + 34 μl	= 552.5 μl

However the seeds are added directly into each well, not into the master mix.

Fig. 6.10 Example calculation to prepare master mix for 17 wells

$Abs_{280\,nm}$ x 10 = Neat $Abs_{280\,nm}$

Neat $Abs_{280\,nm}$ ÷ 2.7 = neat conc (in mg/mL)

10 µg / neat conc (in mg/mL) = volume in L to add per well (typically 20-40 µl)

Volume to add per well x no. of wells = volume of recPrP to add to each master mix.

Fig. 6.11 Example calculation to determine the amount of PrP used per well

*Water volume = 1700 µl (i.e. 17 wells each at 100 µl final volume) – **552.5** (step 1) – PrP volume*

Fig. 6.12 Example calculation to determine the amount of water used per well

4. Perform the following calculation to determine the volume of recPrP solution to be added to each well (Fig. 6.11)—**but do not add the PrP to the master mix yet**:

   Calculate how much water to add by performing the calculation (Fig. 6.12) then add the water to the master mix. The seeds are added into the individual plate wells, once the water and PrP has been added to the master mix—however, the seed volume must still be accounted for at this stage when calculating how much water to add to the master mix (Fig. 6.12).
5. Add the water and mix the master mix well by vortexing or pipetting up and down vigorously.
6. Add the calculated volume of PrP to the master mix.
7. Gently mix by slow inversion or gentle pipetting.

3.4 Assembling the RT-QuIC Plate

Load the RT-QuIC plate using the master mixes (Fig. 6.13). Once the 96-well plate has been loaded, seal the top of the plate using adhesive film and load it into the plate reader (Fig. 6.13).

3.5 Running the RT-QuIC Plate

3.5.1 Summary of RT-QuIC Instrument Protocol

Each "measurement cycle" is comprised of two phases, shaking and reading. These phases are of different length depending on the instrument used. A batch script is used to call the shake and read scripts. Full details and examples of how these are written are listed on the following pages.

Briefly, the FLUOstar Optima and Omega phases are 14 min of shaking followed by 1 min of reading the plate. The 14 min of shaking is programmed in as 7× 2 min shaking cycles on the

Control Master Mix (98 µl / well) then add the control samples (2 µl)
Or
Sample Master Mix (100 µl - CSF sample volume) then add the test CSFs (e.g. 30 µl or 15 µl)

Fig. 6.13 Loading order for the RT-QuIC plate

instrument. The Optima and Omega take slightly different times to accelerate up to the specified shaking speed. This difference results in the Optima having a 2-min cycle consisting of 60 s of shaking and 60 s of rest compared to the Omega's 86 s of shaking and 34 s of rest.

It can take up to 80 h for a sample to give a clear positive result so the measurement cycles are continued until the instrument has completed measurement cycle 361 at around 96 h.

The RT-QuIC assay requires the FLUOstar OMEGA to shake at 900 rpm, which is above the maximum shake speed available through standard software options. However, the manufacturers can modify the software settings such as to allow this higher shake speed to be reached.

It should be noted that the high sensitivity and specificity values reported for RT-QuIC were obtained by using the recommended settings for the FLUOStar Optima instrument. The FLUOstar Omega have been tested in the NCJDRSU and delivered comparable results to the Optima. The use of the recommended conditions is supported by a large body of sample data and deviations from this protocol may lower the quoted sensitivity and specificity values for detection of sCJD.

Some laboratories may only have access to certain instruments, substrates, or a limited access to CSF reference samples. A range of conditions for different instruments and substrates has also been included (Fig. 6.14).

N.B. It is possible to run the RT-QuIC assay without the "Real-Time (RT)" aspect. These conditions rely on two instruments—a separate incubator-shaker for the heating and shaking, then a plate reader for measuring the ThT fluorescence in the wells. Although cheaper to run, it requires the operator to manually perform each plate reading. Commonly, there is a decrease in ThT fluorescence over the course of the experiment and, if only one fluorescent reading is performed at the end of the assay, positive wells may be missed under these conditions.

† The CSF volume used in the assay is a balance between the need for a robust and sensitive assay, the available amount of patient sample and the time taken to complete the measurement. The assay has a sensitivity of 91% and a specificity of 98% when 15µL of CSF is used as per the recommended FLUOstar Optima settings [20].

	Recommended Instrument and Conditions	Alternate Conditions:		Older Instrument and Conditions:	Alternate Conditions with older instrument:			Read only Instrument
Reader	OMEGA	OMEGA		OPTIMA	OPTIMA			TECAN
Recombinant PrP	Ham FL	Ham FL		Ham FL	Ham-Sh Chimeric	Ham FL	Hum FL	Hum FL
CSF volume‡	30 or 15 µl	20 µl	30 µl	15 µl	5 µl	30 µl	30 µl	5 µl
Shake conditions	900rpm 86s shake/34s rest	900rpm 90s shake/30s rest		600rpm 60s shake/60s rest	600rpm 60s shake/60s rest	750rpm 30s shake/30s rest	750rpm 30s shake/30s rest	Max 30s shake/30s rest
Temperature	42°C	42°C		42°C	42°C	37°C	37°C	37°C
Criteria for positive result‡	Mean of 2 highest replicates out of 4 >25,000 rfu at 90 hrs	Mean of 2 highest replicates out of 4 >20,000 rfu at 90 hrs	Mean of 2 highest replicates out of 4 >10,000 rfu at 90 hrs	Mean of 2 highest replicates out of 4 >10,000 rfu at 80 hrs	Mean of 2 highest replicates out of 4 >6,000 rfu at 90 hrs	Mean of 2 highest replicates out of 4 >70% of baseline reading at 90 hrs	Average reading of all four wells >4,344 rfu at 90 hrs	At least 2 of 6 replicates >400 rfu at X hrs

Fig. 6.14 The recommended (and alternative) assay instrumentation and reporting conditions

‡ The rfu "cut-off" value used to determine whether an unknown sample is positive or negative is the mean endpoint of a population of known negative samples + 3 standard deviations. The rfu value also varies with the "gain" setting on the instrument, which is a control of the detector sensitivity. Two different laboratories may therefore use the same instrument and reference samples but use a different rfu "cut-off" limit.

Batch Script for Running an RT-QuIC on either the FLUOstar Optima or Omega (Fig. 6.15)

Shake Script Parameters, RT-QuIC FLUOstar OMEGA (Fig. 6.16)

Read Script Parameters, RT-QuIC FLUOstar OMEGA (Fig. 6.17)

Read Script Parameters, RT-QuIC FLUOstar OPTIMA (Fig. 6.18)

Shake Script Parameters, RT-QuIC FLUOstar OPTIMA (Fig. 6.19)

3.6 RT-QuIC Data Analysis

Samples or controls are positive if **the two wells with the highest rfu** exceed +3 standard deviations of the negative controls within the duration of the run, or at cycle 361. The number of positive wells out of each quadruplicate sample set is noted, as well as the time taken to reach the positive cut-off rfu value (Fig. 6.20).

```
Ask " Have You changed the Expt id" No: Halt Yes:
 st1:="NAME OF EXPERIMENT"                          ; expt id here!
   ID1:="<st1>"                          ; use this value as plate identifier

R_Temp 0.1                               ;switch on temperature monitoring


   TargetTemp:=42.0                       ;set target temperature to 42 degrees celsius

   R_Temp TargetTemp                      ;switch incubator on

   wait for temp >= TargetTemp              ;wait until target temperature reached

Ask " Reader is at 42oC. Insert plate and click Yes to proceed with testrun." No: Halt Yes: R_PlateIn

d:=400                              ;sets loop for 400 times, every 15 minutes for 100 hours

for c:=1 to d do begin                   ;outer loop for multiple readings

     R_Run "RT-QUIC SHAKE"

wait for 60s

     R_Run "RT-QUIC READ"

 ;merge horizontal (kinetic):
   Call "MergeReadings.exe <DataPath> <User> H ID1"


End
```

Fig. 6.15 Batch Script for running RT-QuIC on a FLUOStar plate reader

Out of the brain homogenate controls, the unseeded, sudden-death, and Alzheimer's disease controls should give a flat response. Only the sCJD controls should exhibit an increase in rfu.

Positive CSF samples frequently give between 2/4 and 4/4 positive results.

If only one well out of the four is positive, or if the fluorescent trace rises above the background but does not reach the cut-off value, this could either be:

1. A negative CSF sample exhibiting a stochastic effect due to local well conditions, a "false positive."
2. A positive CSF sample where there are lots of PrP^{Sc} seeds in the wells. In this case, all the recPrP is templated onto many small plaques before the plaques can bind enough ThT to give a strong positive signal.

Shake Script Parameters, RT-QuIC FLUOstar OMEGA:

Basic Parameters:	
Microplate	Nunc 96
Optic	Top Optic
No. of Multichromatics	1
Positioning delay	0.0 (Flying mode unchecked)
No. of kinetic windows	1
Filters	Excitation 450 and Emission filter 480
Gain	2000
Kinetic window 1:	
No. of cycles	7
Measurement start time	0
No. of flashed per well and cycle	0
Cycle time	120
Basic Parameters continued:	
Orbital averaging	(Unchecked)
Minimum cycle time 1	(Blank)
Pause before plate reading	(Unchecked)

Layout:	
Content	Sample
Groups	(Unchecked)
Start value	1, Increase
Replicates	1, Horizontal
Reading direction	Snaking

Concentrations / Volumes / Shaking:	
Standard concentration	Factor (greyed out)
Start volume	0
Volume	Factor, 1
Shaking options	Double orbital, 900 rpm
Additional shaking	Before each cycle
Shaking time	86s

Fig. 6.16 Omega SHAKE script. *N.B. Shaking at 900 rpm requires software update from BMG Labtech*

Read Script Parameters, RT-QuIC FLUOstar OMEGA:

Basic Parameters:	
Microplate	Nunc 96
Optic	Bottom Optic
No. of Multichromatics	1
Positioning delay	0.2 (Flying mode unchecked)
Filters	Excitation filter 450 and Emission filter 480
Gain	2000
Measurement start time	0
No. of flashed per well	20
Orbital averaging	(Unchecked)
Pause before plate reading	(Unchecked)

Layout:	
Groups	(Unchecked)
Start value	1, Increase
Replicates	1, Horizontal
Reading direction	Snaking

Concentrations / Volumes / Shaking:	
Standard concentration	Factor (greyed out)
Start volume	0
Volume	Factor, 1
Shaking options	Double orbital, 900 rpm
Additional shaking	No shaking

Fig. 6.17 Omega READ script

Shake Script Parameters, RT-QuIC FLUOstar OPTIMA:

Basic Parameters:	
Microplate	Nunc 96
Optic	Top Optic
No. of Multichromatics	1
Positioning delay	0.0 (Flying mode unchecked)
No. of kinetic windows	1
Filters	Excitation 450 and Emission filter 480
Gain	2000
Kinetic window 1:	
No. of cycles	7
Measurement start time	0
No. of flashed per well and cycle	0
Cycle time	120
Basic Parameters continued:	
Orbital averaging	(Unchecked)
Minimum cycle time 1	(Greyed out)
Pause before plate reading	0

Layout:	
Content	Sample
Groups	(Unchecked)
Start value	1, Increase
Replicates	1, Horizontal
Reading direction	Snaking

Concentrations / Volumes / Shaking:	
Standard concentration	Factor (greyed out)
Start volume	0
Volume	Factor, 1
Shaking options	Double orbital, shaking width 1mm, 600 rpm
Additional shaking	Before each cycle
Shaking time	57s

Fig. 6.18 Optima SHAKE script

Read Script Parameters, RT-QuIC FLUOstar OPTIMA:

Basic Parameters:	
Microplate	Nunc 96
Optic	Bottom Optic
No. of Multichromatics	1
Positioning delay	0.2 (Flying mode unchecked)
Filters	Excitation filter 450 and Emission filter 480
Gain	2000
Measurement start time	0
No. of flashed per well	20
Orbital averaging	(Unchecked)
Pause before plate reading	(Unchecked)

Layout:	
Groups	(Unchecked)
Start value	1, Increase
Replicates	1, Horizontal
Reading direction	Snaking

Concentrations / Volumes / Shaking:	
Standard concentration	Factor (greyed out)
Start volume	0
Volume	Factor, 1
Shaking options	Double orbital, shaking width 4mm, 150 rpm
Additional shaking	No shaking

Fig. 6.19 Optima READ script

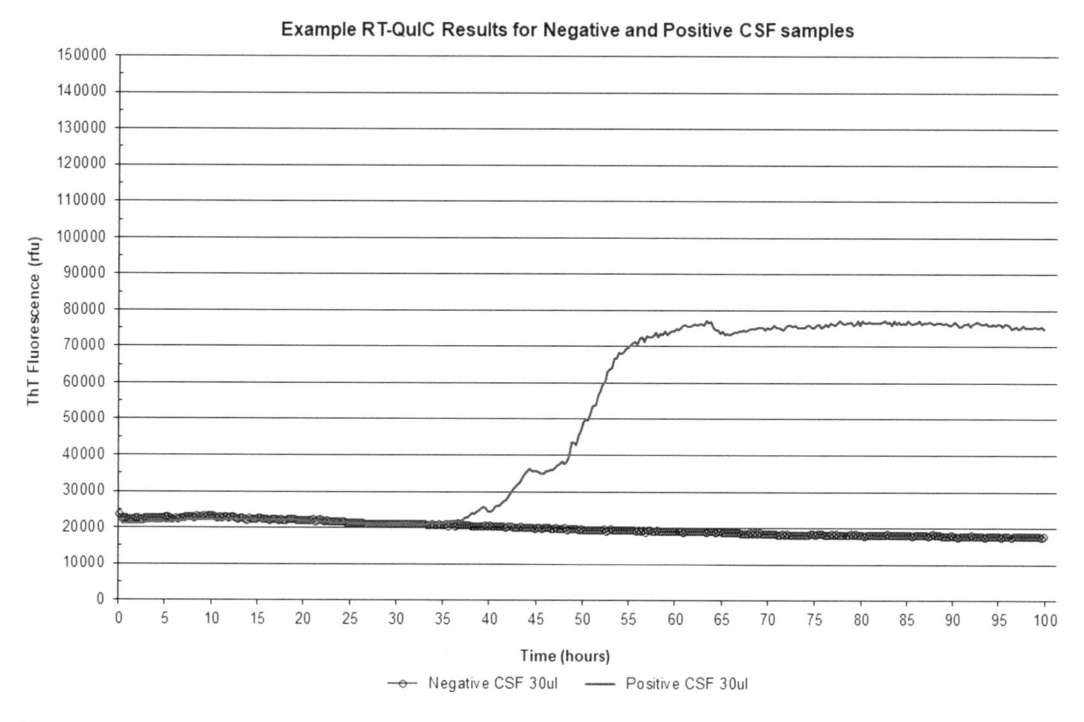

Fig. 6.20 Example positive and negative CSF RT-QuIC traces

In cases where only one well out of the four is positive, the sample analysis is repeated in a subsequent run using 15μL of CSF.

A sample which gives a negative response at 30 and 15μL can give a positive response at 7μL. **Repeating analyses using volumes of CSF <15μL does not give a reliable indication of positivity, based on comparison to pathological examination of *post-mortem* brain.**

Lag Times and End RFU Values in CSF RT-QuIC (Figs. 6.21 and 6.22)

4 Notes

4.1 Troubleshooting RT-QuIC

Frequently Encountered Problems

- **Out of the quadruplicate samples, only 1 or 2 out of the four wells are positive at 30μL sample volume**

  Solution—Repeat the test at both 30μL and 15μL sample volumes. If 3/8 of the wells are positive on the repeat, the sample is positive. If 2/8 or less are positive on the repeat test, then mark the test result as equivocal.

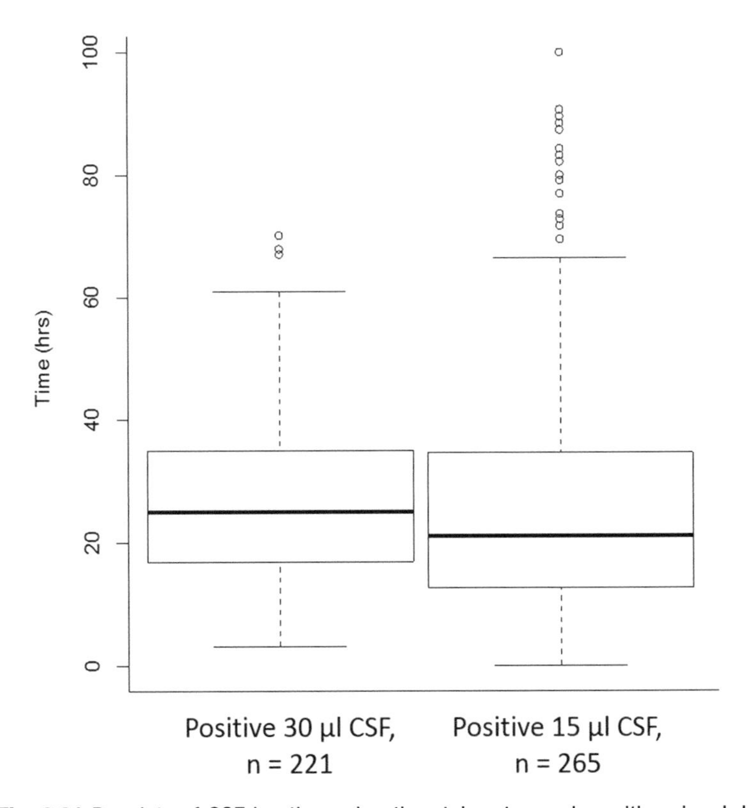

Fig. 6.21 Boxplots of CSF lag times, i.e. time taken to reach positive signal. In 30µL CSFs, Mean = 27.37 h, 1 SD = 13.28 h, 98.7% samples positive by 61 h. In For 15µL CSFs, Mean = 29.10 h, 1 SD = 19.69 h, 98.7% samples positive by 87 h

- **SF collected from a confirmed CJD case does not give a signal at 30µL CSF sample volume**

  Solution—Repeat the test at both 30µL and 15µL sample volumes. If still negative, it may be a genetic CJD or variant CJD case.
- **The rfu baseline is above the positive reporting threshold for the duration of the assay**

  Solution—CSF may have a high protein content or high red/white cell count, run again at 15µL sample volume. If the baseline rfu remains high, look for whether there is a sigmoidal increase in rfu over time. If so, report as positive. If not, report as negative.

  An upper limit for red blood cell count is <1250 red blood cells/µL [18]–however <150 is ideal.

CSFs with red blood cell counts >1250 can give false negatives in RT-QuIC—see Fig. 6.23

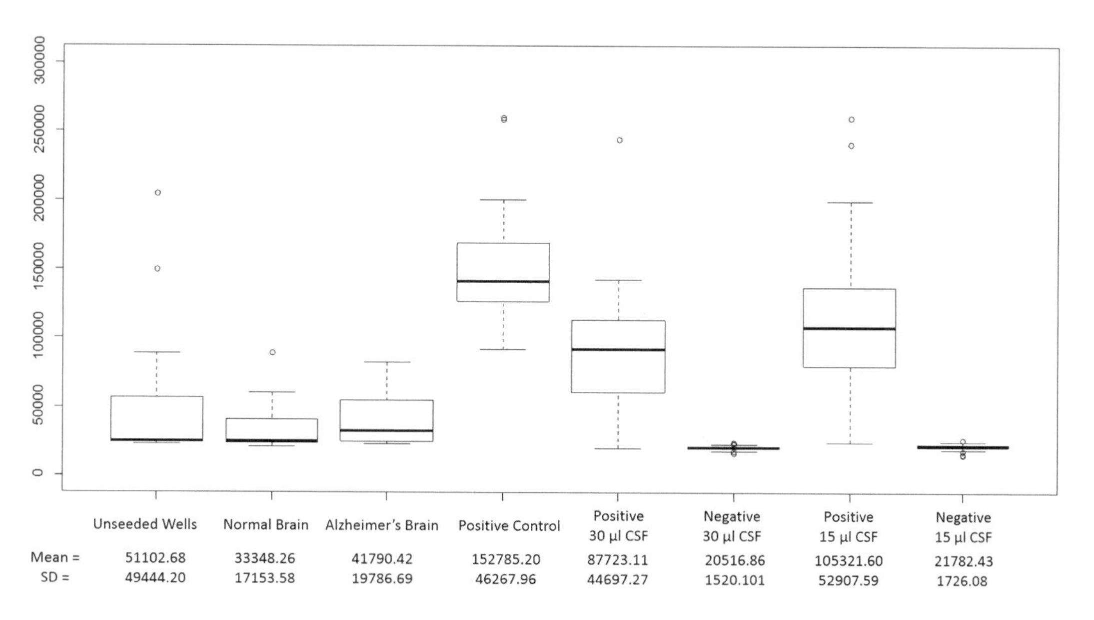

Fig. 6.22 Boxplots of CSF's Endpoint RFU values. In RT-QuIC, if shaken for long enough, all wells will eventually give a positive signal. Occasionally 1 well out of 4 may convert in the negative controls, between 70 and 90 h, and it happens more frequently as the batch of PrP ages. These false positive controls are easily discriminated from true positive CSFs by the lag time, the number of wells which convert (typically $\geq$2 out of 4 wells for a true positive) and the intensity of the RFU rise. Controls samples ($n = 30$); 30μL CSF = (41 positive, 70 negative, $n = 111$); 15μL CSF (106 positive, 89 negative, $n = 195$)

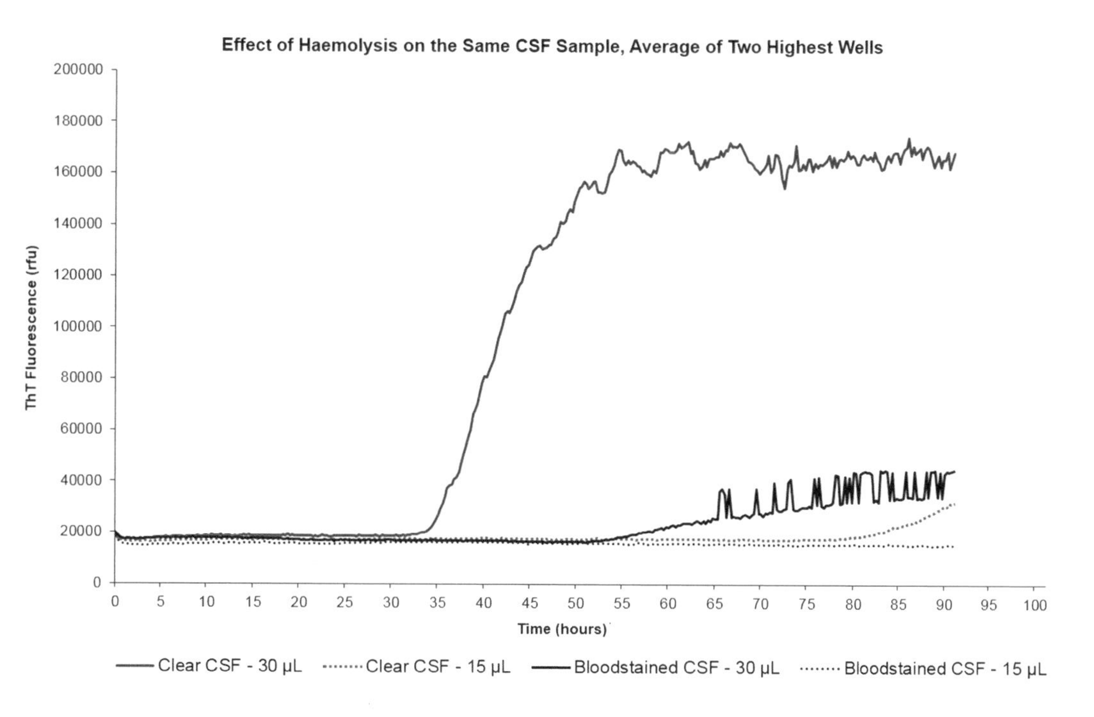

Fig. 6.23 Effect of haemolysis on CSF RT-QuIC

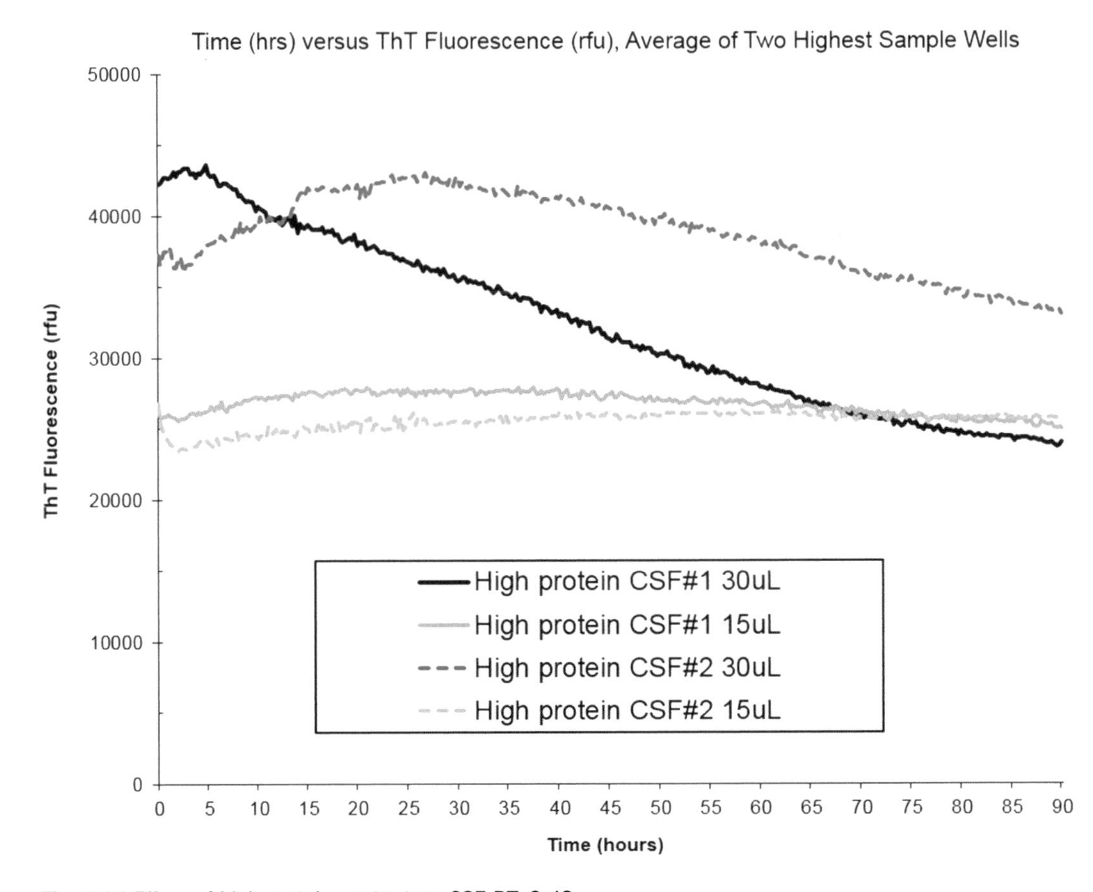

Fig. 6.24 Effect of high protein content on CSF RT-QuIC

- **The rfu baseline has an undulating character**

  Solution—CSF may have a high protein content, run again at 15μL sample volume. If the baseline rfu signal undulates again, look for whether there is a clear sigmoidal increase in rfu over time. If so, report as positive. If not, report as negative.
- **The rfu baseline drifts up over time, gives false positives**

  Solution—this becomes more and more common in PrP batches over 6 months old, despite whether they were stored at −80 °C.

Common Pitfalls of this Method

- Not diluting the PrP before dialysis leads to fibrillization and a massive drop in yield.
- CSF samples contaminated with >1250 red blood cells/mm^3 (e.g., *from traumatic lumbar puncture)* supress the fibrillization of the PrP^c and give a false-negative response (Fig. 6.23)
- CSF samples which are sCJD negative but have a very high protein level can give an RT-QuIC signal with a high or undulating baseline (Fig. 6.24)

Advantages of this Method

1. Compared to other laboratory biomarkers (14-3-3, S100, pTau, and total Tau protein) and compared to clinical investigations (MRI, EEG), the RT-QuIC assay has very high sensitivity (91%) and specificity (98%) for sCJD. It is also sensitive to some genetic forms of CJD [10, 19, 20]
2. It can be used in the differential diagnosis of variant CJD.
3. Once the reagent stocks have been prepared, preparing the assay plate can be completed relatively quickly (1–2 h).
4. Once the assay has been set running, it does not require operator intervention until the assay endpoint.

5 Introduction

5.1 Making Recombinant PrP—Overview

Production of recombinant Hamster Prion protein (Ha recPrP) uses standard techniques in molecular biology and affinity purification. The protein is expressed in BL21 *E. coli* bacteria, purified over a Nickel-IMAC column and is finally dialyzed and concentrated. There are several steps during the purification which are critical to obtaining satisfactory yield, purity, and function. These have been marked at the appropriate point in the procedural sections.

The protocol described here for expression of Ha recPrP uses the BL21 Rosetta DE3/pRSET system from Novagen. Use of the specified bacterium (BL21 *E. coli*) and plasmid (*pRSET*) provides good quality recPrP at high yield.

6 Materials

Recombinant PrP—Consumables (Fig. 6.25)

Recombinant PrP—Equipment (Fig. 6.26)

7 Methods

7.1 Making Recombinant PrP—Molecular Biology

7.2 Disclaimer

When using kits or equipment supplied by third party vendors, please check the supplied documentation as some usage protocols may differ from the typical protocols stated in this chapter.

Material	Amount
Acrylamide gels	1 per purification
Antibiotic powder - e.g. ampicillin or carbenicillin, chloramphenicol	5 g
BL21 DE3 Rosetta E. coli competent cells	1 x 50 uL aliquot
BugBuster Master Mix	0.5 L
Dry Ice or Liquid Nitrogen	As required
Ethanol	1 L
Glycerol, sterile	100 ml
Guanidine.HCl	830 g
Hydrochloric Acid	1 L
Imidazole.HCl	15 g
Isopropanol	0.5 L
Lysogeny Broth 'LB' medium	0.1 L
Molecular weight markers for protein gels	1 per purification
NTA Superflow Ni charged resin	0.5 L
Overnight Express instant TB medium ('OETB' media)	1 x 60 g
pRSET vector containing the gene for Ha PrP	1 x miniprep
Sodium Chloride	9 g
Sodium Hydroxide	As required
Sodium phosphate monobasic	40 g
Sodium phosphate dibasic	40 g
Tris	3 g
Water Ice	As required
Water - distilled or demineralised, sterile.	As required

Fig. 6.25 Overview of the consumables required for producing recombinant PrP

Reagent	Amount
+ 4 oC Fridge or cold room	1
0.22 µ Filter papers for degassing units and filtering solutions	As required
0.22 µ Filters with Luer lock syringe attachments	As required
-20 oC Freezer	1
-80 oC Freezer	1
A packable FPLC column with >150 ml internal volume, e.g. XK 26/40 column	1
Amicon Ultra15 centrifugal units	4
Autoclave tape, 25 M	1
Benchtop rocker / shaker / roller	1
Centrifuge and rotor that can spin 6 x 50 ml Falcon tubes at 13,500 x RCF	1
Containers for liquids - Wet Ice, Dry Ice, Liquid Nitrogen and Dialysis	As required
Cuvette(s) with UV-Vis transparency at 250 nm-700 nm	As required
Dialysis tubing clips	As required
Dialysis tubing with a 7 KDa MWCO	1 roll
Disposable culture flasks, 2.5L, non-baffled	4
Eppendorfs, 1-2 ml.	As required
Equipment for running and visualising SDS-PAGE	As required
Falcon Tubes (50 ml and 15 ml)	As required
Filter unit for degassing buffers	1
FPLC equipment including computer and control software	1
Glass bottles for storing sterile buffers, with capacities from 250 ml – 2 L	As required
Incubator shaker with capacity to shake 4 x 2.5L flasks at 30 oC	1
Measuring cylinders, capacities from 20 ml to 2 L	As required
Metal foil	As required
Microbiological Safety Cabinet Level 2	1
Microfuge	1
Pipettes and sterile tips, ranging from 0.1 µl to 1 ml	As required
Plastic pipettes	As required
Sharpsafe	As required
Spatulas	1
Spectrophotometer, e.g. Nanodrop	1
Syringes, polypropylene Luer lock tips, capacities from 1 ml - 20 mL,	As required

Fig. 6.26 Overview of the equipment required for producing recombinant PrP

7.3 Waste Handling

All material containing genetic material or prion protein should be disposed of in accordance with local laws and health and safety regulations.

7.4 Gene Sequence

The cDNA used to express the Ha recPrP is based on PRNP from *Mesocricetus auratus* (Golden Syrian Hamster), NCBI Gene ID: 101829062. Residues 23-231 are expressed.

The cDNA sequence used for expression is:

5′-aagaagcggccaaagcctggagggtgga
acactggcggaagccgataccctgggcagggcagccctggaggcaaccgttacccacct
cagggtggcggcacatggggacagccccatggtggtggctggggtcagccccatggt
ggtggctggggtcaaccccatggtggtggctggggtcagccccatggtggtggct
ggggtcagggaggtggcacccacaatcagtggaacaagcccagt
aagccaaaaaccaacatgaagcacatggccggcgctgctgcggcaggggccgtggt
gggggccttggtggctacatgctggggagtgccatgagcaggcccatgatgcatttt
ggcaatgactgggaggaccgctactaccgtgaaaacatgaaccgctaccctaaccaagt
gtattaccggccagtggaccagtacaacaaccagaacaactttgtgcacgattgtgt
caacatcaccatcaagcagcacacagtcaccaccaccaccaagggggagaactt
cacggagaccgacatcaagataatggagcgcgtggtggagcagatgtgtaccacccagt
atcagaaggagtcccaggcctactacgatggaagaaggtccagc-3′

Which translates to the protein sequence:

Amino acid residues 23-231:
KKRPKPGGWNTGGSRYPGQGSPGGNRYPPQGGG
TWGQPHGGWGQPHGGGWGQPHGGGWG
QPHGGGWGQGGGTHNQWNKPSKPKTNMKHMAG
AAAAGAVVGGLGGYMLGSAMSRPMMHFGNDWED
RYYRENMNRYPNQVYYRPVDQYNNQNNFVHD
CVNITIKQHTVTTTTKGENFTETD
IKIMERVVEQMCTTQYQKESQAYYDGRRSS

Using standard molecular biology techniques, the above gene sequence should be cloned into a pRSET vector then transformed into BL21 (DE3) pLysS *E. coli* bacteria. The transformed bacteria can then be used to prepare glycerol stocks. Detailed protocols for all of these common steps can be found in "Molecular Cloning—A Laboratory Manual" by Sambrook and Russell, *ISBN 978-1-936113-42-2*.t

7.5 Making Recombinant PrP–Protein Purification

Culturing Bacteria

1. **Preparation of 1000× antibiotic "working stock" solutions for bacterial selection** (0.5 h)
   (a) Dissolve ampicillin (50 mg/mL, 1.00 g) **or** carbenicillin (50 mg/mL, 1.00 g) in sterile water (20 mL).

(b) Dissolve chloramphenicol (25 mg/mL, 0.5 g) in ethanol (20 mL).

(c) Filter the stocks through a 0.22μ filter, then aliquot out into 1 mL volumes.

(d) Store these stocks at −20 °C.

2. **Prepare a Glycerol Stock using Transformed Bacteria** (1 h + overnight incubation)

   Glycerol stocks are a stable way of storing transformed cell lines for several years. They can also be used to inoculate large volumes of media and thus skirt the additional time constraint of producing a fresh starter culture every time a recombinant protein needs to be expressed for purification.

   (a) Obtain fresh and sterile LB medium (≥50 mL) and allow it to come to room temperature.

   (b) Turn on a shaking incubator and set it for 30 °C, 80 rpm.

   (c) In an MBSC Class II, take 4 × 50 mL Falcon tubes labeled A-D and, using filtered aspirator pipettes, dispense LB medium (10 mL) into each one. This will provide two controls and two starter cultures. A—control for aseptic technique, B—control that the bacteria grow without antibiotic present, C, D—starter cultures.

   (d) Add ampicillin and chloramphenicol (i.e., *10μL of each "working stock" solution, ×1000 dilution)* to the LB medium (10 mL) in Falcon tubes C and D.

   (e) Remove the glycerol stock from the −80 °C freezer on dry ice, ensuring it does not thaw during transport or handling.

   (f) Taking 3× sterile toothpicks or pipette tips in hand, dip the narrow end into the glycerol stock and add one tip to Falcon tubes B–D.

   (g) Screw the caps onto the 4× Falcon tubes.

   (h) Transfer the Falcon tubes into the incubator, making sure that they are securely mounted in a vertical orientation within the incubator, i.e., *with lids closest to the ceiling.*

   (i) Grow the cultures up overnight at 37 °C and 180 rpm.

Overnight Incubation Step

(a) In the morning, tube A should be clear, B-D should be cloudy.

- *If A is cloudy then, there is a problem with the operator's aseptic handling technique.*
- *If B is clear, the bacteria do not grow and there is either a problem with bacterial handling or the medium.*

- *If C and D are clear, then there is either a problem with the bacterial transformation step or in preparation of the antibiotic working solution.*

(b) In a 50 mL Falcon tube, prepare an aqueous solution of sterile glycerol solution by adding glycerol (10 mL) to sterile water (10 mL) and vortexing to afford a clear homogenous solution.

(c) Add glucose (0.1 g, 0.5% w/v) to each tube.

(d) Transfer Falcon tubes C and D into an MBSC Class II and, using filter tipped pipettes, dispense 10 mL of the glycerol solution into each tube.

(e) Carefully mix the glycerol solution with the cultures by inverting and rolling the capped tubed several times.

(f) Using filter tipped pipettes and Eppendorfs/vials, dispense the bacterial/glycerol mixes from tubes C and D out into 1x 1 ml aliquots.

(g) Flash freeze the aliquots in liquid nitrogen or by using an acetone/dry ice bath.

(h) Store these glycerol stocks at −80 °C and do not allow them to thaw.

Inoculate the “Grow-up” Cultures (0.5 h)

(a) Transfer 2× non-baffled 2.5 L sterile plastic flasks to an MBSC Class II hood.

(b) To each flask add OETB medium (60 g/L), sterile glycerol (10 ml/L), sterile water (1.0 L), and antibiotics—ampicillin (1 mL working stock/L) and chloramphenicol (1 mL working stock/L).

(c) Dissolve the component by putting the two flasks into an incubator set to 30 °C and 180 rpm. Once dissolved, return the flasks to the MBSC II hood.

(d) Take a further 2× non-baffled 2.5 L sterile plastic flasks into the hood and divide the cultures so that each 2.5 L flask contains 0.5 L volume of liquid.

(e) Use 1–3 mL of glycerol stock culture to inoculate each of these 4 × 500 ml OETB cultures.

(f) Grow the cultures at 30 °C, 180 rpm for 22–26 h.

The OETB media mix does not require manual IPTG induction by the operator. Cultures should be yellow in color and opaque before moving onto the next steps. Around 80–150 mg recPrP is expected from 2 L culture of BL21 DE3 cells.

Pelleting of Bacterial Cells (3 h)

(a) Once the OETB culture is complete, transfer the 4x culture plastic flasks to an MBSC Class II hood.

(b) Taking 6× 50 mL Falcon tubes, fill each tube with culture up to the 40 mL mark and sort them into three pairs.

(c) Balance the tube pairs to within 10 mg on a balance.

(d) Pellet the bacterial cells at 3000 × RCF for 15 min.

(e) Discard the supernatant to waste, retaining the cell pellets in the Falcon tubes.

(f) Decant further portions of culture into the Falcon tubes, balance pairs to within 10 mg then repeat the centrifugation step until all of the culture has been pelleted and supernatant waste has been discarded.

(g) The Falcon tubes are convenient storage containers for subsequent lysis steps.

These "whole cell" pellets may be stored at this stage for up to 6 months at −80 °C.

N.B. Expected yield is around 2 g of cell pellet per Falcon tube when the freezing step has been reached.

Preparation of Purification Buffers (Expect 4–6 h)

The purification can be performed at 18–20 °C without detriment to the protein; however, the dialysis step should be at 0–4 °C. The water used to dilute the dialysis buffer does not have to be de-gased. Prepare the buffers as described below. The pH tolerances are ±0.05 units. Filter the buffers through a 0.22μ filter and de-gas prior to use.

7.6 "8DB" Denaturing Buffer

Dissolve guanidine hydrochloride (76.42 g, 800 mmol, MW = 95.53 g/mol) in demineralized water (30 mL).

N.B. This dissolution is very endothermic. 8 M is a concentration near the solubility limit for guanidine hydrochloride. The water volume will increase by a large amount as the salt dissolves.

*Adjust final volume to **0.10 L at pH 8.00.***

Final guanidine concentration = 8.00 mol/L.

7.7 "6DB" Denaturing Buffer

Dissolve sodium phosphate monobasic (1.20 g, 10 mmol, MW = 119.98 g/mol), sodium phosphate dibasic (12.78 g, 90 mmol, MW = 141.96 g/mol), tris (1.21 g, 10 mmol, MW = 121.14 g/mol), and guanidine hydrochloride (573.18 g, 6 mol, MW = 95.53 g/mol) in demineralized water (300 mL).

*Adjust final volume to **1.00 L at pH 8.00.***

Final guanidine concentration = 6.00 mol/L.

Final phosphate concentration = 100 mmol/L.

Final tris concentration = 10 mmol/L.

7.8 "RB" Refolding Buffer

Dissolve sodium phosphate monobasic (1.20 g, 10 mmol, MW = 119.98 g/mol), sodium phosphate dibasic (12.78 g,

90 mmol, MW = 141.96 g/mol), and tris (1.21 g, 10 mmol, MW = 121.14 g/mol) in demineralized water (950 mL).

*Adjust final volume to **1.00 L at pH 8.00.***
Final phosphate concentration = 100 mmol/L.
Final tris concentration = 10 mmol/L.

7.9 "EB" Elution Buffer

Dissolve sodium phosphate monobasic (2.40 g, 20 mmol, MW = 119.98 g/mol), sodium phosphate dibasic (0.71 g, 5 mmol, MW = 141.96 g/mol), tris (0.30 g, 2.5 mmol, MW = 121.14 g/mol), and imidazole hydrochloride (13.07 g, 125 mmol, MW = 104.54 g/mol) in demineralized water (200 mL).

*Adjust final volume to **0.25 L at pH 6.00.***
Final imidazole concentration = 500 mmol/L.
Final phosphate concentration = 100 mmol/L.
Final tris concentration = 10 mmol/L.

7.10 Cleaning Resin—"DEB" Denaturing Elution Buffer

Dissolve sodium phosphate monobasic (3.60 g, 30 mmol, MW = 119.98 g/mol), sodium chloride (8.77 g, 150 mmol, MW = 58.44 g/mol), and guanidine hydrochloride (171.95 g, 1.8 mol, MW = 95.53 g/mol) in demineralized water (100 mL).

*Adjust final volume to **0.30 L at pH 4.00.***
Final guanidine concentration = 6.00 mol/L.
Final phosphate concentration = 100 mmol/L.
Final NaCl concentration = 500 mmol/L.

7.11 "20× DIB" (20× Dialysis Buffer)

Dissolve sodium phosphate monobasic (20.4 g, 170 mmol, MW = 119.98 g/mol) and sodium phosphate dibasic (4.0 g, 30 mmol, MW = 141.96 g/mol) in demineralized water (0.95 L).

*Adjust final volume to **1.00 L at pH 5.8***
Final phosphate concentration = 10 mmol/L.
After 20× dilution, the pH should be pH 6.5.

7.12 NaOH Solution

N.B. this is only used for adjusting the pH of buffers and is not required for recPrP purification.

Dissolve sodium hydroxide (60.0 g, 1.5 mol, MW = 40.0 g/mol) in demineralized water (0.25 L).

Adjust final volume to **0.30 L at ca. pH 14.7.**
Final hydroxide concentration = 5.00 mol/L.

Isolating the Inclusion Bodies

1. **Double Freeze-Thawing of Cells** (2 h)
   (a) Remove the "whole cell" pellets from the −80 °C freezer and allow them to warm up over 15–30 min at room

temperature. *To save time, this first thaw/freeze step can be performed on the day before the purification.*

(b) Return the pellets to the −80 °C freezer and allow them to freeze (15–30 min).

(c) Remove the "whole cell" pellets from the −80 °C freezer and keep them on ice.

2. **Isolate the Inclusion Bodies** (1.5 h)

(a) Remove BugBuster Master Mix from the fridge ca. 30 min before use and allow it to warm up to room temperature. *N.B. Make sure that it is a homogenous solution before use, as it contains detergent which can drop out of solution during storage in the fridge.*

(b) Using a pastette, resuspend the "whole cell" pellets in Bug Buster Master Mix. Expect to use around 15 mL/2 g of cell pellet and do not vortex.

(c) Gently agitate the contents for 30 min at room temperature on, e.g., a benchtop rocker at 10 rpm or horizontally mounted on a shaking incubator at 70 rpm. After this incubation step, the content should be a homogenous suspension of inclusion bodies and cell pellet debris.

(d) Pellet the inclusion bodies by balancing pairs of tubes to within 10 mg of each other, then spinning the pairs at 13,500 × RCF for 10 min at 4 °C.

(e) Keep the inclusion body pellets in the bottom of the Falcon tubes and discard the supernatant to waste.

3. **First Wash of the Inclusion Bodies** (2 h)

(a) Resuspend the inclusion body pellets in Bug Buster Master Mix (15 mL).

(b) Gently agitate the contents for 30 min, at room temperature on, e.g., a benchtop rocker at 10 rpm or horizontally mounted on a shaking incubator at 70 rpm.

(c) Pellet the inclusion bodies by spinning at 13,500 × RCF for 10 min at 4 °C.

(d) Keep the pellets and discard the supernatant to waste.

4. **Second Wash of the Inclusion Bodies** (2 h)

(a) Take Bug Buster Master Mix (10 mL) and dilute with distilled water (90 mL) to afford a 10% Bug Buster solution.

(b) Resuspend the inclusion bodies and debris in this solution (10 mL/per pellet).

(c) Gently agitate the contents for 30 min at room temperature on, e.g., a benchtop rocker at 10 rpm or horizontally mounted on a shaking incubator at 70 rpm.

(d) Pellet the inclusion bodies by spinning at 13,500 × RCF for 10 min at 4 °C.

(e) Keep the pellets and discard the supernatant to waste.

After isolation and washing, the inclusion bodies should be milky-white. When frozen, they should be a dark yellow-green in color.

At this point in the protocol, the washed inclusion bodies can be frozen at −80 °C and stored for up to 6 months.

Preparation of the Dialysis Buffer (0.3 h Preparation, ca. 3 h Chilling)

(a) Take "20×" dialysis buffer (0.2 L) and dilute with distilled water (3.8 L) to afford a 1× solution.

(b) Adjust the buffer pH to 5.8 as required.

(c) Chill the buffer to 4 °C in preparation for dialysis.

Washing of the Ni NTA Resin (2 h—*The Equilibration is Comprised of Three Buffer Washes.*)

(a) To 6 × 50 mL Falcon tubes add Nickel NTA Superflow slurry (35 mL of the 50% slurry; 17 mL of resin after settling).

(b) To the resin in each tube add 6 M Guanidine denaturing buffer ("6DB," 10 mL) then equilibrate the resin using a benchtop shaker or rocker (5–10 min, room temperature).

(c) Spin at 1500 × RCF for 5 min to sediment the resin.

(d) Wash the resin in each tube with 10 mL of the 6 M Guanidine denaturing buffer ("6DB") using a rocker at 10 rpm for 5 min.

(e) Spin at 1500 × RCF for 5 min to sediment the resin.

(f) Wash the resin in each tube with 10 mL of the 6 M Guanidine denaturing buffer ("6DB") using a rocker at 10 rpm for 5 min.

(g) Spin at 1500 × RCF for 5 min to sediment the resin.

(h) Decant off the supernatant and replace with enough "6DB" to afford a final slurry volume of 20 mL per tube.

Loading recPrP onto Ni NTA Resin (3 h)

1. **Lysis of the inclusion bodies**

   (a) Resuspend each inclusion body pellet in 8 M denaturing buffer ("8DB," 15 mL).

   (b) Pellet the debris for 10 min at 4 °C and 13,500 × RCF.

   (c) Add the supernatants (6 × ca. 15 mL) to the equilibrated matrix slurries in the above step.

   (d) Resuspend the debris in a further 8 M denaturing buffer ("8DB," 15 mL).

   (e) Pellet the debris for 10 min at 4 °C and 13,500 × RCF.

   (f) Add the supernatants (6 × ca. 15 mL) to the equilibrated matrix slurries in the above step.

2. **Load lysate onto the Nickel NTA resin**
   (a) Mix the inclusion body lysates with the equilibrated NTA resin (in 6 × Falcon tubes).
   (b) Shake on a rocker at 10 rpm for 60 min at 18–20 °C.
   (c) Spin down the matrix at 1500 × RCF for 5 min at 18–20 °C then discard the supernatants to waste.
3. **Purge the FPLC system**
   Equilibrate the FPLC system with 6 M Denaturing Buffer "6DB" (Line A) and refolding buffer "RB" (Line B).

Purifying the recPrP (ca. 6 h but an overnight "hold point" is possible during the refolding step.)

1. **Pack the column with the Nickel NTA resin** *(15 min)*
   (a) Pour the equilibrated NTA resin slurry into an XK26/60 column (or another column with an internal capacity of ≥100 mL) and bed the matrix.
   (b) Connect the column to the FPLC.
2. **Wash non-PrP protein off the column** *(25 min)*
   Pump isocratic 6 M denaturing buffer "6DB" (5 min at 5 mL/min, ca. 0.25 column volumes) across the column.
3. **Refold recPrP on the column** *(3 h 35 min)*
   (a) Run a gradient switching from the current 6 M denaturing buffer "6DB" (line A) to the renaturing buffer "RB" (line B) (5 mL/min for 200 min, ca. 10 column volumes).
   The refolding gradient can be stopped at 50–80% of B and left overnight at room temperature without substantial detriment to the final protein quality or the equipment.
   (b) Once the gradient is complete, continue pumping isocratic renaturing buffer "RB" (line B) over the column (5 mL/min for 15 min, ca. 0.75 column volumes).
4. **Elute recPrP** *(1 h 20 min)*

(a) Once the refolding step is complete, prepare the system to run an elution gradient, switching from refolding buffer (RB) to elution buffer (EB).
Remember to purge the pump and line of "6DB" buffer before starting the elution gradient, or a small volume of denaturing buffer will be pumped onto the column.

(b) Start the elution gradient (5 mL/min for 60 min, ca. 3 column volumes), switching from the current renaturing buffer "RB" to the elution buffer "EB."

(c) Once the UV Abs trace rises, begin collecting the eluate in 10 mL fractions.

Observations:

The PrP peak will elute when the buffer gradient is comprised of 50% elution buffer and ca. 100 mL of eluate will be collected. The peak should be broad (ca. 30 min) with little to no front shoulder. The UV trace will reach a maximum of 1000 mAU.

PrP from the front and back of the elution peak does not work well in RT-QuIC.

Good quality PrP is found in the fractions where the UV abs rises above 450 mAU.

(a) Identify which collected fractions had UV abs >450 mAU.

(b) Combine these fractions and dilute them into 1 x dialysis buffer at a ratio of volume eluate to 2 volumes of buffer.

(c) Filter this solution through a 0.22μ filter—this increases final recPrP yield by reducing loss to spontaneous fibrillization.

(d) Once the gradient is complete, continue pumping the isocratic elution buffer at 5 mL/min over the column for a further 20 min (ca. 1 column volume).

The Abs_{280nm} baseline will rise slowly during the elution step since imidazole absorbs at 280 nm. The elution endpoint is at the conclusion of this isocratic step. The earliest and latest eluting fractions are unsuitable for use in RT-QuIC.

Dialysis 1 *Overnight (8–14 h)*

(a) Prepare another 4 L of fresh 1× dialysis buffer as specified earlier, so that it is prechilled and ready for the next day's buffer exchange.

(b) Transfer the filtered and diluted PrP fractions into dialysis tubing (7 kDa MWCO), place in prechilled 1× dialysis buffer (4 L, 2–8 °C, pH 5.8) and incubate overnight with stirring.

During the dialysis steps, some PrP will fibrillize. This will be visible in the dialysis tubing as short white threads around 1–2 mm in length. Extending the dialysis leads to further protein loss and will negatively affect yield, but not final reactivity in the RT-QuIC assay.

Dialysis 2 *Exchange the Dialysis Buffer (4–5 h)*

(a) Discharge the 1× dialysis buffer from the first dialysis step to waste.

(b) Exchange the dialysis buffer for fresh, prechilled 1× dialysis buffer and continue dialysis for a further 4–5 h.

Determine recPrP Concentration

Determine the dialyzed recPrP's concentration using UV-Vis spectroscopy or BCA assay kits.

As an example, using UV-Vis spectroscopy:

(a) Prepare a solution of 0.01% SDS in 1× PBS by diluting 10% SDS (100μL) into 1× PBS (10 mL) and filtering through a 0.22μ filter.

(b) Dilute the concentrated PrP solution (100μL) into the PBS/SDS mixture (900μL).

(c) Measure the absorption at 280 nm.

$$\text{Concentration} = \text{Abs}_{280\ \text{nm}}/(\varepsilon_{280\ \text{nm}} \times \text{cell path length in cm})$$

For Ha PrP FL:

$\varepsilon_{280\text{nm}} = 62{,}005\ \text{M}^{-1}\ \text{cm}^{-1}$ (1 g/L = 2.702)

MW = 23.0603 kDa.

Concentrate recPrP (3–4 h)

(a) Using centrifugal concentrators in accordance with the manufacturer's instructions, concentrate the PrP to give a final concentration of between 0.3 and 0.5 mg/mL.

(b) Aliquot the concentrated PrP into 1 mL fractions using 1.5–2 mL Eppendorfs. Retain 1 × 1 mL for quality control in the next steps.

(c) Flash-freeze the PrP by immersing the Eppendorfs in liquid nitrogen or a dry ice/acetone ice bath then store at −80 °C until required.

Quality Control

(a) Determine the purity of the concentrated recPrP using, e.g., SDS-PAGE and Western blot.

If substantially contaminated, then do not proceed to the next step.

(b) RecPrP quality is determined by testing in RT-QuIC against the control brain homogenates and known positive and negative CSF samples. The recPrP batch passes if it produces the expected signal responses as set out in the data analysis section.

A "double overlapping band" for recPrP is often observed by SDS-PAGE when purifying with this method but it has no detrimental effect on the RT-QuIC assay.

Regenerating the Ni NTA Resin

1. **Washing the Resin**
   (a) Switch both FPLC lines to denaturing elution buffer ("DEB").
   (b) Pump the denaturing elution buffer at 5 mL/min over the column for 30 min (ca. 1.5 column volumes).
   (c) Switch both FPLC lines to water.
   (d) Pump the water at 5 mL/min over the column for 30 min (ca. 1.5 column volumes).
2. **Recondition the Resin**
   (a) Transfer the resin from the column into 6 × 50 mL Falcon tubes (ca. 17 mL resin/tube).
   (b) Add demineralized water (15 mL) to each tube and spin the resin down at 3000 RCF for 5 min.
   (c) Wash 1—Decant off the supernatant and wash with **0.5 M NaOH** (15 mL).
   (d) Spin the resin down at 3000 RCF for 5 min.
   (e) Wash 2—Decant off the supernatant and wash with **0.5 M NaOH** (15 mL).
   (f) Spin the resin down at 3000 RCF for 5 min.
   (g) Wash 3—Decant off the supernatant and wash with **0.5 M NaOH** (15 mL).
   (h) Spin the resin down at 3000 RCF for 5 min.

N.B. A color change of blue to brown may be observed during this step.

(a) Wash 4—Decant off the supernatant and wash with **demineralized water** (15 mL).
(b) Spin the resin down at 3000 RCF for 5 min.
(c) Wash 5—Decant off the supernatant and wash with **demineralized water** (15 mL).
(d) Spin the resin down at 3000 RCF for 5 min.
(e) Wash 6—Decant off the supernatant and wash with **renaturing buffer** ("RB," 15 mL).
(f) Spin the resin down at 3000 RCF for 5 min.
(g) Wash 7—Decant off the supernatant and wash with **renaturing buffer** ("RB," 15 mL).
(h) Spin the resin down at 3000 RCF for 5 min then decant off the supernatant.

N.B. If storing for a 1–2 weeks, store the matrix in renaturing buffer at 4 °C. If storing for longer, use a 20% ethanol/water solution at 4 °C.

8 Notes

8.1 Troubleshooting PrP Production

Frequently Encountered Problems

- **Low yield of PrP**

  Solution—There are multiple possible causes as follows:

  1. Make sure that the glycerol stocks used to inoculate the cultures have not thawed and been refrozen. This kills cells within the glycerol stock and lengthens the required culture time.
  2. Cells did not grow as expected—try lengthening the duration of the bacterial culture step by 2 h.
  3. Make sure that during the resolubilization of the whole cell and inclusion body pellets, all pelleted material has been completely solubilized.
  4. If the purification buffers were made up and filtered in advance, check each buffer pH is within the tolerance limits on the day of purification.
  5. Perform 0.2μ filtrations and dilutions at every indicated step in the protocol.
  6. Check the molecular weight cut-off limit for the dialysis membranes, whether they require treatment before use and whether they are sealed correctly before starting the dialysis step.
  7. Check the manufacturer's instructions for spin concentrators, checking that the RCF limit for the membrane is not exceeded and that the molecular weight cut-off limit for the centrifugal membrane is smaller than the molecular weight of the protein.

- **Side peak on front of main PrP peak during the elution**

  Using the conditions described, the peak maximum should be around 1000 mAU; however the PrP should only be collected when the UV absorption has risen above 450 mAU. If the side peak is particularly large then,

  Solution

  1. Improve washing steps to remove contaminating proteins by making sure the pellets are wholly resuspended, or repeat the inclusion body pellet washing steps.
  2. Alternatively, the column could be overloaded, try using more resin.

- **Spontaneous fibrillization**

  Some fibrillization is normally visible at the end of the dialysis step and is removed by the 0.2μ filtration step, without detriment to final yield. If fibrillization is excessive and adversely impacts the yield,

  Solution

  1. Perform 0.2μ filtrations and dilutions at every indicated step in the protocol.
  2. Check buffer pHs are correct.
  3. Check buffers were filtered correctly, there is no dust or rough surfaces on the dialysis container which could be acting as nucleation points for the recPrP.

Common Pitfalls of this Method

- Not filtering solutions at the indicated steps leads to fibrillization and a massive drop in yield.
- Incomplete freeze-thawing or Bug Buster lysis of the whole cell pellets leads to impure isolations of inclusion bodies. This can cause impurities in the final product or a drop in yield.
- Poor pH control leading to weak binding of the PrP to the resin and a massive drop in yield.

Advantages of this Method

- After the initial cost outlay for equipment and training, in house production of PrP allows for fine control over protein quality along with low production cost and abundant protein yields.

References

1. Will RG, Alperovitch A, Poser S, Pocchiari M, Hofman A et al (1998) Descriptive epidemiology of Creutzfeldt-Jakob disease in six European countries, 1993-1995. Ann Neurol 43:763–767
2. Zerr I, Kallenberg K, Summers DM, Romero C, Taratuto A et al (2009) Updated clinical diagnostic criteria for sporadic Creutzfeldt-Jakob disease. Brain 132:2659–2668
3. World Health Organisation. 1998. Global surveillance, Diagnosis and therapy of human transmissible spongiform encephalopathies: Report of a WHO Consultation. Online
4. Saborio GP, Permanne B, Soto C (2001) Sensitive detection of pathological prion protein by cyclic amplification of protein misfolding. Nature 411:810–813
5. Peden AH, McGuire LI, Appleford NEJ, Mallinson G, Wilham JM et al (2012) Sensitive and specific detection of sporadic Creutzfeldt-Jakob disease brain prion protein using real-time quaking-induced conversion. J Gen Virol 93:438–449
6. Atarashi R, Satoh K, Sano K, Fuse T, Yamaguchi N et al (2011) Ultrasensitive human prion detection in cerebrospinal fluid by real-time quaking-induced conversion. Nat Med 17:175–178
7. Wilham JM, Orru CD, Bessen RA, Atarashi R, Sano K et al (2010) Rapid end-point quantitation of prion seeding activity with sensitivity comparable to bioassays. PLoS Pathog 6

8. Jarrett JT, Lansbury PT (1993) Seeding one-dimensional crystallization of amyloid - a pathogenic mechanism in alzheimers-disease and scrapie. Cell 73:1055–1058
9. Orro CD, Groveman BR, Hughson AG, Manca M, Raymond LD et al (2017) RT-QuIC assays for prion disease detection and diagnostics. Methods in Molecular Biology (Clifton, NJ) 1658:185–203
10. Lattanzio F, Abu-Rumeileh S, Franceschini A, Kai H, Amore G et al (2017) Prion-specific and surrogate CSF biomarkers in Creutzfeldt-Jakob disease: diagnostic accuracy in relation to molecular subtypes and analysis of neuropathological correlates of p-tau and A beta 42 levels. Acta Neuropathol 133:559–578
11. Groveman BR, Orru CD, Hughson AG, Bongianni M, Fiorini M et al (2017) Extended and direct evaluation of RT-QuIC assays for Creutzfeldt-Jakob disease diagnosis. Annals of Clinical and Translational Neurology 4:139–144
12. Foutz A, Appleby BS, Hamlin C, Liu X, Yang S et al (2017) Diagnostic and prognostic value of human prion detection in cerebrospinal fluid. Ann Neurol 81:79–92
13. Bongianni M, Orru C, Groveman BR, Sacchetto L, Fiorini M et al (2017) Diagnosis of human prion disease using real-time quaking-induced conversion testing of olfactory mucosa and cerebrospinal fluid samples. JAMA Neurol 74:155–162
14. Park J-H, Choi Y-G, Lee Y-J, Park S-J, Choi H-S et al (2016) Real-time quaking-induced conversion analysis for the diagnosis of sporadic Creutzfeldt-Jakob disease in Korea. J Clin Neurol 12:101–106
15. McGuire LI, Poleggi A, Poggiolini I, Suardi S, Grznarova K et al (2016) Cerebrospinal fluid real-time quaking-induced conversion is a robust and reliable test for sporadic Creutzfeldt-Jakob disease: an international study. Ann Neurol 80:160–165
16. Orru CD, Groveman BR, Hughson AG, Zanusso G, Coulthart MB, Caughey B (2015) Rapid and sensitive RT-QuIC detection of human Creutzfeldt-Jakob disease using cerebrospinal fluid. mBio 6
17. Cramm M, Schmitz M, Karch A, Zafar S, Varges D et al (2015) Characteristic CSF prion seeding efficiency in humans with prion diseases. Mol Neurobiol 51:396–405
18. Cramm M, Schmitz M, Karch A, Mitrova E, Kuhn F et al (2016) Stability and reproducibility underscore utility of RT-QuIC for diagnosis of Creutzfeldt-Jakob disease. Mol Neurobiol 53:1896–1904
19. Sano K, Satoh K, Atarashi R, Takashima H, Iwasaki Y et al (2013) Early detection of abnormal prion protein in genetic human prion diseases now possible using real-time QUIC assay. PLoS One 8
20. McGuire LI, Peden AH, Orrú CD, Wilham JM, Appleford NE et al (2012) Real time quaking-induced conversion analysis of cerebrospinal fluid in sporadic Creutzfeldt–Jakob disease. Ann Neurol 72:278–285

Chapter 7

Lumbar Puncture: Consensus Guidelines

Pim Fijneman and Sebastiaan Engelborghs

Abstract

Cerebrospinal fluid collection by lumbar puncture (LP) is performed in the diagnostic workup of several neurological brain diseases. Reluctance to perform the procedure is among others due to a lack of standards and guidelines to minimize the risk of complications, such as post-LP headache or back pain.

We provide a summary of consensus guidelines for the LP procedure to minimize the risk of complications. Our consensus guidelines provide a step-by-step procedure description and address contraindications, as well as patient-related and procedure-related risk factors that can influence the development of post-LP complications. When an LP is performed correctly, the procedure is well tolerated and accepted with a low complication rate.

Key words Lumbar puncture, Cerebrospinal fluid, Post-LP complications, Headache, Back pain, Consensus guidelines, Evidence-based guidelines

1 Introduction

Lumbar puncture (LP) is a technique to sample cerebrospinal fluid (CSF), which is a window to brain pathology. It involves the introduction a needle into the subarachnoid space of the lumbar sac, at a level safely below the spinal cord [1]. Despite modern neuro-imaging techniques, LP remains an important diagnostic tool as CSF analysis provides important diagnostic information for many neurological conditions. No procedure can replace the CSF analysis in differential diagnosis of infectious disorders of the central nervous system like bacterial or viral meningitis and neuroborreliosis. Moreover, CSF analysis is now at the core of the diagnostic criteria for the diagnosis of Alzheimer's disease [2–4]. In addition, an LP is the easiest procedure to perform a CSF pressure measurement.

A large international, multicenter study on LP feasibility that included 3868 patients in a memory clinic setting, showed that LPs can be safely performed [5]. The acceptance rate of an LP was high, especially taking into consideration that there was no acute medical

Charlotte E. Teunissen and Henrik Zetterberg (eds.), *Cerebrospinal Fluid Biomarkers*, Neuromethods, vol. 168, https://doi.org/10.1007/978-1-0716-1319-1_7,

indication. The most important independent risk factors for post-LP complaints were related to patient characteristics: history of headache and fear of complications. Younger age is the most important and well-known risk factor for PLPH and post-LP back pain [5]. It has been reported that PLPH is more common in women than in men, and especially women below 40 years of age seem to have a substantially higher risk of PLPH, whereas cognitive impairment or dementia is associated with a risk reduction of PLPH [6]. A cutting bevel needle type appeared to be the main procedure-related risk factor for typical PLPH. The number of LP attempts was the main procedure-related risk factor for occurrence of local back pain. A large needle diameter ($\leq$22G, gauge (G)) was a risk factor for severe headache [5].

In order to reduce complication rates, evidence-based guidelines have been published that are summarized in this chapter [6].

2 Contraindications

Before the LP procedure is started, contraindications must be ruled out. The most important contraindications for LP are an intracranial space-occupying lesion with mass effect, a posterior fossa mass, an increase of intracranial pressure due to increased CSF pressure or an Arnold-Chiari malformation, which can lead to cerebellar or tonsillar herniation [6]. Other contraindications include suspected spinal epidural abscess, coagulopathies and uncorrected bleeding diathesis, use of anticoagulant medication, congenital spine abnormalities, and local skin infection over the needle entry site.

A brain computed tomography (CT) or magnetic resonance imaging (MRI) scan should at least be performed before an LP if a intracranial lesion with mass effect or abnormal intracranial pressure is suspected, or whenever a patient has focal neurological deficits, impaired consciousness, recent seizures, previous central nervous system disease, papilledema at fundoscopy or is immune compromised. Evidence for a pressure gradient across the falx cerebri (risk of uncal herniation) or between the supra- and infratentorial compartments (risk of uncal herniation) or Arnold-Chiari malformation (risk of tonsillar herniation) on brain imaging should be considered as a contraindication for an LP.

Brain or spinal hemorrhage and spinal epidural or subdural hematoma are very rare but serious complications of an LP. Therefore, it is advised to have a recent blood analysis with a platelet count above 40×10^9/L and a normal coagulation status (Quick >50%; International Normalized Ratio (INR) < 1.5) [6]. Anticoagulant medication should be stopped to minimize the hemorrhagic risks. This decision, however, must be based on the risk of the discontinuation itself (i.e., increased risk of thrombosis), the possibility of bridging with heparin as well as the potential advantage of the LP and possible alternative examinations. Antiplatelet drugs do not need to be interrupted, however in the case of dual antiplatelet therapy (e.g., clopidogrel and acetylsalicylic acid) it is advised for a nonurgent LP, to temporarily withhold (1–2 weeks) the intake of thienopyridine derivatives (e.g., clopidogrel) unless patients are at a high thrombotic risk.

In case of spine abnormalities, the LP can be guided with imaging (e.g., ultrasound) to visualize the patient-specific anatomical variations. Performing an LP through infected skin reduces the diagnostic value in case of a suspected central nervous system infection; it as well increases the risk of central nervous system infections.

3 LP Procedure

3.1 Equipment

All required material should be gathered before the procedure. This should include:

- sterile gloves
- sterile drapes
- sterile gauze dressing
- antiseptic solution
- LP needle
- polypropylene collection tubes
- wound dressing

Chlorhexidine, alcohol-based solutions, and povidone iodine can all be used as skin disinfectant.

3.2 Needle Selection

The selection of the optimal LP needle (length, diameter, and design) should be based on the medical indication, adequate flow rate, and CSF pressure measurement, but foremost it should minimize the discomfort for the patient and risk of complications. Regular length needles (70–90 mm) are suitable for most patients, long spinal needles (>90 mm) are only required for some obese patients. The LP-feasibility study performed in memory clinic settings did not show that large bore diameter needles confer independent risk compared to small bore needle types [5]. However, smaller-bore needles (defined as ≥24G) are recommended by the majority of studies in the literature [6], based on lower incidence of PLPH, back pain, and discomfort. A lower incidence of complications resulted in less medical assistance and less medication after the procedure. The recommended needle design is atraumatic as opposed to cutting bevel needles. Atraumatic needles have a lower frequency of complications (consequently lower health care costs) and traumatic taps. The disadvantages are that these needles require a more technical procedure (and thus require more training) and can lead to more attempts and failures.

3.3 Patient Positioning (Lateral Recumbent Position or Seated)

To start the procedure, the patient must be adequately positioned either in the sitting position or the lateral recumbent position. In the lateral recumbent position (left-sided for right-handed physicians), the patient should assume a fetal position with their back in flexion and knees pulled up towards their chest. The hips, legs, and shoulders should be parallel to each other and perpendicular to the floor. In the upright seated position, the patient should flex their neck and back completely forward, while embracing a pillow to maximize lumbar flexion. Keep shoulders in alignment to prevent twisting of the spine. Both positions are intended to overcome the lumbar lordosis and therefore facilitating the course of the needle by widening the gap between adjacent lumbar spinal processes.

The preferred position depends on the preference of the physician and on the patient. In case of ill or immobilized patients, the lateral recumbent position should be used. For CSF pressure measurement, patients need to be in the lateral recumbent position. In the sitting position, there is a higher CSF pressure and flow and thus a shorter procedure time; however, this position is associated with an increased risk of severe headache, albeit not for typical PLPH [5–7].

3.4 Needle Insertion

The aim is to reach the lumbar cistern inferior to the conus medullaris. In the majority of people, this bottommost part of the spinal cord terminates at L1/L2. In children, the spinal cord ends at L3. To determine the site of insertion, Tuffier's line is often used, which is a virtual line joining the most superior part of both iliac crests. The intersection of this line with the midline of the lumbar spine is the spinal process of L4 of the interspace L4-5. The skin should be disinfected. Needle insertion should be at the L3-4 or L4-5 interspace in adults, in children interspace L4-5 or L5-S1 should be used. The insertion of the needle should be at the superior aspect of the inferior spinous process, while angling the needle towards the umbilicus (15° cephalad). When using a cutting bevel needle, it is preferred to hold the bevel in the sagittal plane in an effort to reduce the damage to the dura mater by separating its longitudinal fibers rather than cutting through them.

The number of LP attempts was significantly associated with post-LP back pain [5]. A total of four attempts is regarded as an acceptable maximum [6]. Passive withdrawal of CSF resulted in a lower risk for nonspecific headache and a strongly lower risk for severe headache [5]. Active withdrawal (syringe) obviously reduces the procedural time, especially with small-bore needles. If active withdrawal is chosen, a higher risk of PLPH should be taken into account [6].

3.5 Measuring CSF Pressure

Depending on the indication for the LP, CSF pressure measurement could be the first step after reaching the correct needle position. A manometer can be attached to the hub of the LP needle. The pressure is expressed in cm H_2O, corresponding to the height of CSF in the tube of the manometer above the puncture site recorded in centimeters. The opening pressure can be disrupted by disorders that dysregulate the production, flow or absorption of CSF, but interpreting the opening pressure should be done with care because patient position or changes in intra-abdominal pressure can also influence the pressure. Small pressure variations can be observed related to normal breathing of the patient, whereas a Valsalva maneuver (e.g., coughing) can lead to bigger spikes in pressure. In the lateral recumbent position, the normal CSF pressure range is 10–18 cm H_2O (adults) and 3–6 cm H_2O in children.

3.6 Volume of CSF to Be Collected

Collection of up to 30 mL of CSF is well tolerated and safe [7] and is advised as an acceptable maximum [6].

4 Conclusions: Recommendations

These recommendations should minimize post-LP complications, the most frequent being PLPH and post-LP back pain [6].

First, contraindications for LP should be ruled out:

- It is advised to perform brain imaging prior to LP, whenever an intracranial lesion with mass effect, abnormal intracranial pressure due to increased CSF pressure, or tonsillar herniation is suspected based on medical history or neurological examination, and in case of recent seizures, impaired consciousness, or papilledema.
- Coagulation status and platelet count (should be above 40×10^9/L) should be checked by (recent) blood analysis prior to LP.
- Concomitant medication should be checked prior to LP. In case of intake of anticoagulants, an LP is contraindicated unless the risk of the procedure outweighs the potential benefit. Direct acting anticoagulants can be temporarily interrupted. An LP can be performed without substantial risk when patients take one type of antiplatelet drug.
- Infections at the LP site are relative contraindications.

As patient-related characteristics are among the most important risk factors for PLPH and post-LP back pain, the physician should determine the risk profile of the patient:

- Younger age, being a female below 40 years of age, previous history of headache, and fear of post-LP complications are risk factors for PLPH and post-LP back pain.
- Post-LP complaints are less prevalent in patients with cognitive deterioration [5].

Recommendations with regard to the LP procedure itself:

- It is recommended to use 25G atraumatic needles given the reduced incidence of PLPH.
- A total of four attempts may be regarded as an acceptable maximum, as the risk for back pain significantly increased with >4 attempts [5, 6].
- Active CSF withdrawal using a syringe should only be performed when a patient cannot tolerate a long procedure. If a large volume of CSF has to be withdrawn (e.g., for research purposes or in case of an evacuating LP), a larger (preferably atraumatic) needle diameter is recommended instead of active withdrawal by using a syringe.
- It is recommended to perform an LP in the lateral recumbent position due to the fact that the sitting position was associated with more severe headache [5–7]. For CSF pressure measurement, patients need to be in the lateral recumbent position.
- The collection of up to 30 mL of CSF is well tolerated and safe [7].

As local anesthesia, and bed rest after LP are not associated with decreased prevalence of post-LP complications, there are no recommendations to apply.

An informative video is a helpful complementary tool to these guidelines; it can be downloaded from the supplemental materials section of Babapour et al., 2017 [8].

In conclusion, an LP is a common and generally well-tolerated diagnostic procedure with a high diagnostic yield. The application of these evidence-based guidelines will help to reduce complication rates.

References

1. Niemantsverdriet E et al (2015) Techniques, contraindications and complications of CSF collection procedures. In: Deisenhammer F, Sellebjerg F, Teunissen CE, Tumani H (eds) Cerebrospinal fluid in clinical neurology. Springer, pp 35–57
2. McKhann GM et al (2011) The diagnosis of dementia due to Alzheimer's disease: recommendations from the National Institute on Aging-Alzheimer's Association workgroups on diagnostic guidelines for Alzheimer's disease. Alzheimers Dement 7:263–269
3. Albert MS et al (2011) The diagnosis of mild cognitive impairment due to Alzheimer's disease: recommendations from the National Institute on Aging-Alzheimer's Association workgroups on diagnostic guidelines for Alzheimer's disease. Alzheimers Dement 7:270–279

4. Dubois B et al (2014) Advancing research diagnostic criteria for Alzheimer's disease: the IWG-2 criteria. Lancet Neurol 13:614–629
5. Duits FH et al (2016) Performance and complications of lumbar puncture in memory clinics: results of the multicenter lumbar puncture feasibility study. Alzheimers Dement 12:154–163
6. Engelborghs S et al (2017) Consensus guidelines for lumbar puncture in patients with neurological diseases. Alzheimers Dement 18:111–126
7. Monserrate AE et al (2015) Factors associated with the onset and persistence of post-lumbar puncture headache. JAMA Neurol 72:325–332
8. Babapour Mofrad R et al (2017) Lumbar puncture in patients with neurologic conditions. Alzheimers Dement 8:108–110

Chapter 8

Pre-Analytical Processing and Biobanking Protocol for CSF Samples

Charlotte E. Teunissen and Eline Willemse

Abstract

This chapter describes the methods for pre-analytical processing and biobanking of cerebrospinal fluid (CSF) samples. There is increasing knowledge that pre-analytical procedures and storage conditions can influence biomarker results, which is relevant to all kinds of fluids and thus also biomarkers measured in CSF. The pre-analytical phase includes patient-related factors, such as fasting, as well as processing factors, such as delay between collection and processing, type of tube, or number of transfers of the fluid. As for the processing factors, a distinction can be made whether analysis is performed for routine care or research purposes, which mostly affects the storage time and storage temperatures. For example, routine analysis usually implies storage of days to weeks at −20 °C or 4 °C, while research analysis usually implies storage for years at −80 °C. The presented protocol is based on previously published protocols for CSF collection and long-term biobanking in the context of biomarker development or research, with adaptations based on experimental evidence collected over time.

Key words Biobanking, Cerebrospinal fluid, Pre-analytical processing, Tube, Transfer, Freezing, Thawing, Storage, Biomarker research

1 Introduction

This chapter describes the methods for pre-analytical processing and biobanking of cerebrospinal fluid (CSF) samples. The application of international standardized and uniform methods for pre-analytical processing and biobanking are of enormous importance for biomarker development, including discovery, to obtain sufficiently large samples sizes, replication in independent cohorts collected under the same conditions, and for clinical implementation, for which the biomarkers usually have to be analyzed in a large variety of independent and large cohorts. The pre-analytical procedures and storage conditions can influence results of biomarkers, which accounts for all kinds of fluids and thus also biomarkers

The chapter therefore provides two different protocols pertaining to these two different purposes.

Charlotte E. Teunissen and Henrik Zetterberg (eds.), *Cerebrospinal Fluid Biomarkers*, Neuromethods, vol. 168,
https://doi.org/10.1007/978-1-0716-1319-1_8,

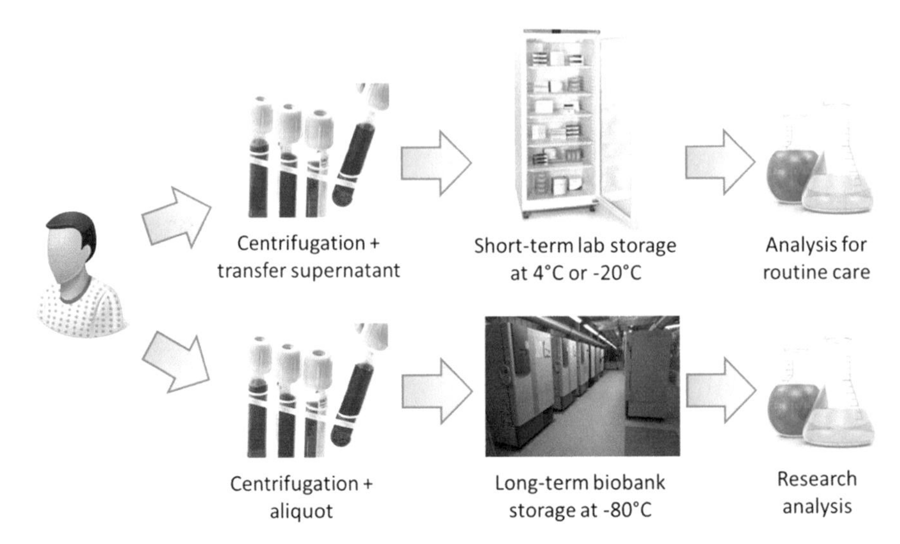

Fig. 8.1 Schematic diagram of (pre-)analytic workflow for routine care biomarker analysis (upper workflow) and for research-related biomarker analysis (bottom workflow)

measured in CSF, even though the markers are not equally affected by all types of variation.

The route from sample collection to finally measurement of the analyte of interest consists of several processing steps and factors which are susceptible to variation (Fig. 8.1). All variation that could occur during collection and processing steps is together called "pre-analytical variation" and can be caused by human biology, e.g. diurnal rhythm in the patient, environmental factors, e.g. temperature in the laboratory, or human handling, e.g. tube transfer. A distinction can be made whether analysis is performed for routine care, which usually means storage of days to weeks at −20 °C or 4 °C and use of patient identifiers, or research analysis, which usually mean storage for years at −80 °C and requires coding of material (Fig. 8.1).

Variation during measurement of the analyte of interest has appointed the term "analytical variation" and can be caused by composition of the assay, quality of the standards and antibodies, machine settings, environment, or human handling. Analytical interferences during the analytical phase will be addressed in other chapters.

In general laboratory medicine, 70% of all diagnostic errors were found to be caused by errors in the pre-analytical phase [1, 2]. This prompted the development of internal quality controls and external quality assessments for general laboratory medicine to reduce this type of errors [3]. Likewise, there is an urgent need to harmonize the pre-analytical variation factors in the laboratory for CSF, as (pre-)analytical variation factors accounted for 10–30% of misdiagnosis of AD upon reanalysis [4]. Since the beginning of this century, biomarkers in CSF have been used to support the diagnosis of Alzheimer's disease. While up till then CSF was only used for routine clinical measurements, such as cell count, the discovery of diagnostic markers in CSF led to collection and storage of CSF samples in biobanks, to use for clinical follow-up and to search for novel biomarkers. With the growth of CSF biobanks, there was a need for consensus guidelines for CSF processing and storage to achieve harmonization of biobank collections across centers [5–8]. Especially for multicenter studies, it is crucial to minimize the effect of (pre-) analytical variation between centers. The presented protocol is based on those previously published protocols for CSF collection and long-term biobanking in the context of biomarker development or research [5–7, 9–13], with adaptations based on experimental evidence collected over recent years. Protocols for specific analytes or technologies can deviate from this more generic protocol, depending on specifics identified for these analytes.

2 Materials

1. Type of needle: Atraumatic, large gauze (smaller size): >23 g, see chapter on lumbar puncture procedures.
2. Collection tubes for CSF: Polypropylene tubes, screw cap, volume >10 mL (**Note 1**).
3. Collection tubes for serum: no clotting activator or gel.
4. Collection tubes for EDTA-plasma: no protease inhibitors.
5. Aliquoting tubes: same for all fluids: polypropylene tubes, volume such that tubes will be filled >50% and preferably >75%. The usual tube volumes are 0.5, 1 and 1.5 mL. *See* **Notes 2** and **3**.
6. Freezing-resistant labels.
7. Centrifuge able to accommodate speed range of 1200 *g* and different temperature settings (4 °C, room temperature).
8. 4 °C degree refrigerator (routine).
9. −20 °C freezer (routine).
10. −80 °C freezer (research/Biobanking).
11. Biobanking information system in order to record, summarize, and update information of each individual sample.

3 Methods

3.1 Biobanking Protocol

Stepwise procedure:

1. Upon arrival of the sample in the lab, check if the appropriate tubes have been received and check labeling of patient/subject information.
2. Centrifuge the sample at the speed indicated in Table 8.1, point 8.

Table 8.1
Collection protocol for CSF and blood pairs for biobanking

Item no	Procedure	Ideal situation for CSF	Blood
A: Collection procedures			
1	Time of day of withdrawal and storage	Record date and time of collection	Same as for CSF
2	Preferred volume	At least 12 mL. First 1–2 mL for routine CSF assessment. Last 10 mL for biobanking Record volume taken and fraction used for biobanking, if applicable	10 mL EDTA-plasma, 10 mL serum
3	Location	Intervertebral space L3-L5(S1)	Venepuncture
4	If blood contamination occurred	Do not process further Criteria for blood contamination: more than 500 red blood cells/μL Record number of blood cells in diagnostic samples	na
5	Other body fluids that should be collected simultaneously	Serum	na
5	Other body fluids that should be collected simultaneously	Plasma: EDTA (preferred over citrate)	na
B. Processing for storage			
7	Storage temperature until freezing	Room temperature before, during, and after centrifugation	Same as CSF
8	Centrifugation conditions	2000 ×*g* (1800–2200), 10 min at room temperature. 400 ×*g* if cells are to be preserved	2000 ×*g* (1800–2200), 10 min at room temperature

(continued)

Table 8.1 (continued)

Item no	Procedure	Ideal situation for CSF	Blood
9	Time delay between withdrawal, processing, and freezing	Between 30 and 60 min. Max 2 h After centrifugation, samples should be aliquoted and frozen immediately, with a maximal delay of 2 h	Between 30 and 60 min. Max 2 h Less than 1 h is optimal for proteomics discovery studies. Serum must clot minimal 30 min
10	Aliquoting	A minimum of two aliquots is recommended. The advised research sample volume of 10 mL should be enough for >10 aliquots	As CSF
11	Volume of aliquots	Minimum 0.1 mL. Depending on total volume of tube: 0.2, 0.5 and 1 mL. Preferably, the tubes are filled up to 75% of the volume. *See* **Note 2**	As CSF
12	Coding	Unique codes. Freezing-proof labels. Ideally barcodes to facilitate searching, to aid in blinding the analysis, and to protect the privacy of patients	As CSF
13	Freezing temperature	−80 °C. *See* **Note 6**	As CSF

3. During centrifugation, insert the subject information into the biobank information management system and print labels for the aliquots. Label the aliquoting tubes.
4. Distribute the supernatant over the aliquoting tubes in standard volumes (e.g., always 0.5 mL and only the last one with the left-over).
5. Freeze immediately at −80 °C. The whole procedure should be finished within 2 h for both CSF and blood products. *See* **Notes 4** and **5**.

4 Notes

1. Polypropylene collection tube should be used for collection of CSF for Biobanking purposes. It is well described that transfer of sample from one tube to the next leads to a loss of amyloid proteins [14], which can be counteracted by the analysis of ratios of amyloid beta(42)/amyloid beta (40) in case of the amyloids, as all amyloid proteins show similar absorption. Of note, absorption occurs already during contact with the pipette

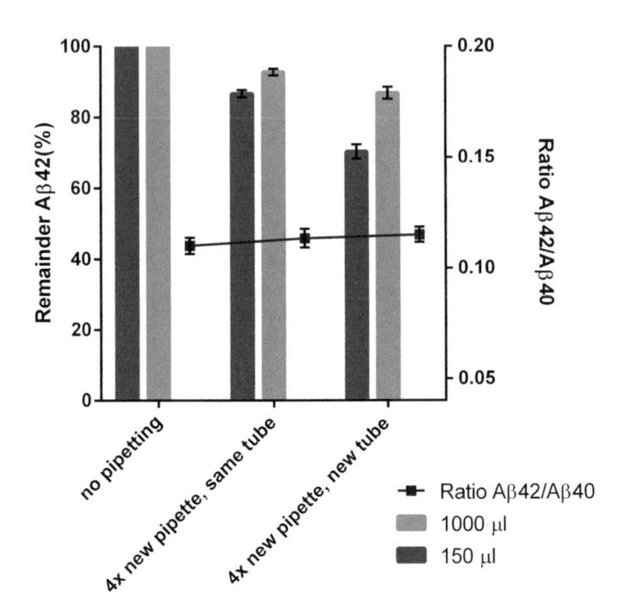

Fig. 8.2 Effect of protein absorption on amyloid beta(1-42) concentration and the ratio amyloid beta(1-42)/amyloid beta(1-40) in CSF during pipetting. (Adapted from [14])

tip, on top of contact with the new aliquoting tubes [14] (Fig. 8.2). Nevertheless, an alternative route could be to avoid transfer and directly insert the collection tube into the analysis automate. A second tube needs than to be used to collect additional volume of CSF for Biobanking and further research.

2. In a previous study, we showed that evaporation does not occur during biobank storage at −20 °C or −80 °C for the (body) fluids CSF, plasma, serum, saliva, or water, for up to 4.5 years and at different percentages of filling of the tube (at least >2.5%, minimal volume tested was 50 μL in a 2 mL tube) [15](Fig. 8.3). However, care should be taken to close the lid carefully.
3. Smaller aliquots are usually optimal to balance between the expected volume needed, which is not yet know at time of biobanking, and the cost of storage, which is of course higher when multiple smaller aliquots are stored. While there is limited knowledge on the difference between direct analysis and one time freezing, for research projects one time freezing for storage cannot be avoided. However, the systematic studies performed on CSF so far indicate that the vast majority of markers no differences are observed in concentrations between 1 time and up to 7 times freezing and thawing [16]. For example, we found that >70% of the proteins remained stable under the most extreme conditions of one-week storage at

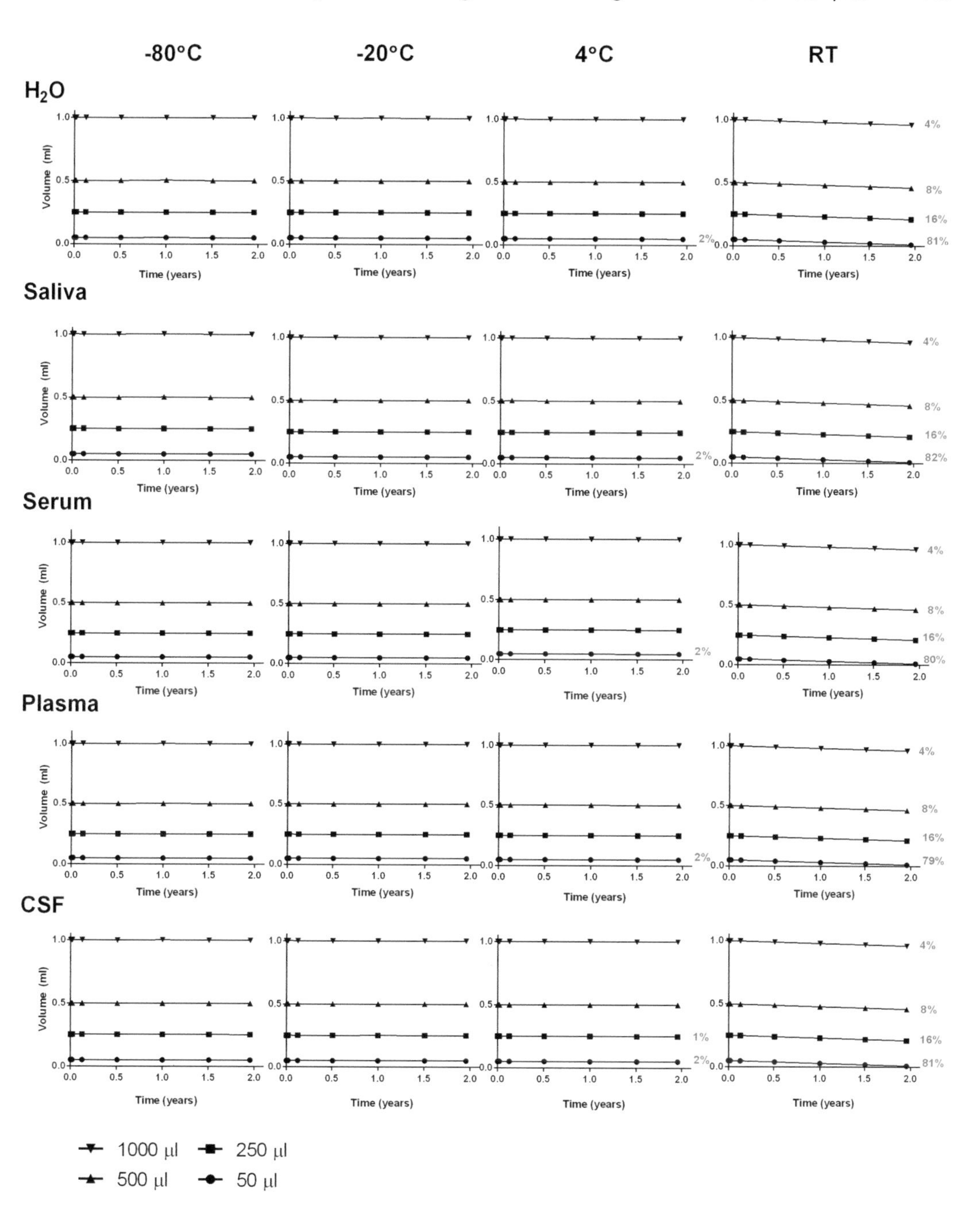

Fig. 8.3 Evaporation of different body fluids and water at different temperatures monitored over 2 years. *RT* = room temperature. (Adapted from [15])

room temperature or after 7 freeze/thaw cycles compared to the reference sample. This indicates that the large majority of CSF proteins remains stable when exposed to time delays between processing steps up to 1 week, or multiple freeze/thaw cycles [11].

4. Even though the recommendation is to finish the procedure within 2 h from collection, there is so far no evidence that longer processing duration has major effects [11].
5. There are currently a couple of tools available to check the quality of the Biobanking procedures. First, the self-assessment tool of International Society for Biological and Environmental Repositories (ISBER) for evaluating the biobanking quality for blood, tissue, and cells has been expanded with questions on the pre-analytical handling of CSF. The self-assessment tool tests the compliance of the biobank to best practices, and thereby stimulates the harmonization of biobank collection, processing, and storage procedures that are the basis to perform reproducible biomarker studies [17]. Secondly, the Integrated Biobank of Luxembourg (IBBL) runs an annual biobank proficiency testing scheme to compare performances of biobank sample work-up, such as aliquoting, amongst laboratories. Proficient biobanks subsequently get certified and accredited [18], which increases the quality of biobank samples and increases reproducibility of biomarker studies. The first round of this program showed differences in biomarker outcomes (amyloid beta(42), hyperphosphorylated tau(181), and albumin were tested) between the processing centers. The results form the basis to assess quality of the Biobanking procedures and adaptations of these procedures [19].
6. Effects of long-term biobanking are hard to study, especially when assays are still under development and can thus not be used to analyze the value at intermittent time-points over longer periods. We designed a study to evaluate the AD CSF biomarker concentrations of amyloid beta(42), tau, and hyperphosphorylated tau(181) stored at −80 °C biobank for up to 12 years [20]. Our study design assumed that the mean biomarker concentrations in a homogeneous AD patient cohort would remain constant on group level over time, thereby allowing measurement of the biomarkers in patient samples with different storage times (2–14 years) while excluding assay batch interference. The biomarker concentrations measured using one assay batch and remained stable regardless of the biobank storage time. Thus, we conclude that long-term biobank storage at −80 °C can be safely applied for the biomarkers amyloid beta(42), tau, and hyperphosphorylated tau (181), for at least up to 12 years.

References

1. Plebani M (2006) Errors in clinical laboratories or errors in laboratory medicine? Clin Chem Lab Med 44(6):750–759. https://doi.org/10.1515/CCLM.2006.123
2. Lippi G, Guidi GC, Mattiuzzi C, Plebani M (2006) Preanalytical variability: the dark side of the moon in laboratory testing. Clin Chem Lab Med 44(4):358–365. https://doi.org/10.1515/CCLM.2006.073

3. Plebani M, Sciacovelli L, Aita A (2017) Quality indicators for the total testing process. Clin Lab Med 37(1):187–205. https://doi.org/10.1016/j.cll.2016.09.015
4. Vos SJB, Visser PJ, Verhey F et al (2014) Variability of CSF Alzheimer's disease biomarkers: implications for clinical practice. PLoS One 9 (6):e100784. https://doi.org/10.1371/journal.pone.0100784
5. del Campo M, Mollenhauer B, Bertolotto A et al (2012) Recommendations to standardize preanalytical confounding factors in Alzheimer's and Parkinson's disease cerebrospinal fluid biomarkers: an update. Biomark Med 6 (4):419–430. https://doi.org/10.2217/bmm.12.46
6. Teunissen CE, Petzold A, Bennett JL et al (2009) A consensus protocol for the standardization of cerebrospinal fluid collection and biobanking. Neurology 73(22):1914–1922. https://doi.org/10.1212/WNL.0b013e3181c47cc2
7. Teunissen CE, Tumani H, Bennett JL et al (2011) Consensus guidelines for CSF and blood biobanking for CNS biomarker studies. Mult Scler Int 2011:246412. https://doi.org/10.1155/2011/246412
8. Vanderstichele H, Bibl M, Engelborghs S et al (2012) Standardization of preanalytical aspects of cerebrospinal fluid biomarker testing for Alzheimer's disease diagnosis: a consensus paper from the Alzheimer's biomarkers standardization initiative. Alzheimers Dement 8 (1):65–73. https://doi.org/10.1016/j.jalz.2011.07.004
9. Vanderstichele HMJ, Janelidze S, Demeyer L et al (2016) Optimized standard operating procedures for the analysis of cerebrospinal fluid Aβ42 and the ratios of Aβ isoforms using low protein binding tubes. J Alzheimers Dis 53:1121–1132. https://doi.org/10.3233/JAD-160286
10. Willemse EAJ, Teunissen CE (2015) Biobanking of cerebrospinal fluid for biomarker analysis in neurological diseases. In: Karimi-Busheri F (ed) Biobanking in the 21st century, advances in experimental medicine and biology. Springer International Publishing, Basel, pp 79–92. https://doi.org/10.1007/978-3-319-20579-3
11. Hok-A-Hin YS, Willemse EAJ, Teunissen CE, Del Campo M (2019) Guidelines for CSF processing and biobanking: impact on the identification and development of optimal CSF protein biomarkers. In: Humana, New York, NY, pp 27–50. https://doi.org/10.1007/978-1-4939-9706-0_2
12. Willemse EAJ, Teunissen CE (2015) Importance of pre-analytical stability for CSF biomarker testing. https://doi.org/10.1007/978-3-319-01225-4_5
13. Teunissen CE, Verheul C, Willemse EAJ (2017) The use of cerebrospinal fluid in biomarker studies. In: Deisenhammer F, Teunissen CE, Tumani H (eds) Handbook of clinical neurology, cerebrospinal fluid in neurologic disorders. Elsevier, pp 3–20. https://doi.org/10.1016/B978-0-12-804279-3.00001-0
14. Willemse E, van Uffelen K, Brix B, Engelborghs S, Vanderstichele H, Teunissen C (2017) How to handle adsorption of cerebrospinal fluid amyloid-β (1–42) in laboratory practice? Identifying problematic handlings and resolving the issue by use of the Aβ 42/Aβ 40 ratio. Alzheimers Dement 13 (8):885–892. https://doi.org/10.1016/j.jalz.2017.01.010
15. Willemse EAJ, Koel-simmelink MJA, Durieux-lu S, Van Der Flier WM, Teunissen CE (2012) Standard biobanking conditions are sufficient to prevent biofluid sample evaporation. 2013
16. Willemse EAJ, Vermeiren Y, Garcia-Ayllon M-S et al (2019) Pre-analytical stability of novel cerebrospinal fluid biomarkers. Clin Chim Acta 497:204–211. https://doi.org/10.1016/J.CCA.2019.07.024
17. Pitt K, Betsou F (2012) The ISBER best practices self assessment tool (SAT): lessons learned after three years of collecting responses. Biopreserv Biobank 10(6):548–549. https://doi.org/10.1089/bio.2012.1064
18. Gaignaux A, Ashton G, Coppola D et al (2016) A biospecimen proficiency testing program for biobank accreditation: four years of experience. Biopreserv Biobank 14(5):429–439. https://doi.org/10.1089/bio.2015.0108
19. Lewczuk P, Gaignaux A, Kofanova O et al (2018) Interlaboratory proficiency processing scheme in CSF aliquoting: implementation and assessment based on biomarkers of Alzheimer's disease. Alzheimers Res Ther 10 (1):87. https://doi.org/10.1186/s13195-018-0418-3
20. Willemse EAJ, van Uffelen KWJ, van der Flier WM, Teunissen CE (2017) Effect of long-term storage in biobanks on cerebrospinal fluid biomarker Aβ 1-42 , T-tau, and P-tau values. Alzheimers Dement Diagn Assess Dis Monit 8:45–50. https://doi.org/10.1016/j.dadm.2017.03.005

Chapter 9

Communicating Complex Results of Cerebrospinal Fluid Analysis

Axel Regeniter and Werner H. Siede

Abstract

Cerebrospinal fluid (CSF) analysis requires a combined assessment of multiple tests to establish diagnosis. Some result combinations require the application of specific rules, based on mathematical formulae to supply a correct result interpretation. This chapter addresses some typical pitfalls in the interpretation of blood contaminated CSF, interpretation of the antibody specifity index, and markers of dementia. The text output of a knowledge-based system combined with a visually oriented report form facilitates the recognition of these typical but often not apparent disease patterns.

Key words CSF evaluation program, Visually oriented knowledge-based system, Visually oriented integrated report form, CSF signature pattern, Disease-related data patterns and evaluation, CSF blood contamination correction, CSF antibody specificity index interpretation, Intrathecal immunoglobulin interpretation, CSF dementia marker interpretation

Abbreviations

ASI	Antibody specificity index
CBC	Complete blood differential
CSF	cerebrospinal fluid
Ec	Erythrocytes
Leuco	Leucocytes
Mono	Mononuclear cells
NMO	Neuromyelitis optica (M. Devic)
OCB	Oligoclonal bands
Poly	Polymorphic cells

1 Introduction

The literature emphasizes the importance of CSF analysis in the workup of acute and chronic neurological diseases. Even the basic workup, however, requires the simultaneous analysis and correct

Charlotte E. Teunissen and Henrik Zetterberg (eds.), *Cerebrospinal Fluid Biomarkers*, Neuromethods, vol. 168,
https://doi.org/10.1007/978-1-0716-1319-1_9,

interpretation of multiple analytes of CSF and serum values, i.e., cell count, lactate, glucose, albumin, Ig profiles, isoelectric focusing, and in some cases, immunofixation. All relate to each other; some must be adjusted to age. CSF diagnostics, however, is based on mathematical formulae. Their correct implementation and interpretation are mandatory to obtain a valid diagnosis. Barrier function and Ig profiles, for instance, have to be evaluated with Reiber's formula and should be additionally plotted on a log/log scale to evaluate border function and intrathecal immunoglobulin production [1]. The analysis of these results is especially valuable because there is only an incomplete or no isotype switch from IgM to IgG synthesis inside the CNS [2, 3]. These CSF changes in basic analysis are diagnostically most relevant, differ substantially and often create a typical signature pattern, for instance, in multiple sclerosis or neuroborreliosis (Fig. 9.1). All the significant diagnostic information will be neglected, however, if the relevant information cannot be interpreted correctly by the treating physician. Sadly, complex lab results on conventional report forms are still presented as long lines of numbers and insufficiently flagged as "high" or "low" value only. Related results are often scattered on several different sheets of paper. The interpretation is left to the ordering clinician in most cases. This can be a difficult and time-consuming task, particularly when the clinician is confronted with laboratory tests outside of his expertise.

We therefore developed a CSF module for our visually oriented knowledge-based system [4, 5] and visualize the most important data (Fig. 9.1). Quantitative measurements are divided by their reference range to remove the effect of age and gender and are shown as bar plots. Reference range values are plotted in blue; elevated values in red. The pattern present in isoelectric focusing forms the central part of the report form because it is the best analytical method to detect intrathecal production which detects an additional 30% of intrathecally produced IgG when compared with nephelometry [5]. The visual summary of the most important analytes in basic CSF analysis creates a characteristic signature pattern for many acute and chronic neurological diseases and is substantiated with text output from the built-in knowledge data base.

In multiple sclerosis (Fig. 9.1a) patients usually present with a reference range or only marginally elevated CSF white blood cell count. The age-dependent CSF-serum quotient for albumin is reference range in 80% of cases or only slightly elevated. Isoelectric focusing shows oligoclonal bands in 95–100% of the patients, whereas the quantitative IgG increase is present only in 70–80% [6, 7].

Lyme disease (borreliosis) has an estimated high incidence in the range of 60,000 to more than 200,000 cases per year in Germany [8]. Acute neuroborreliosis presents in 3.3% of cases

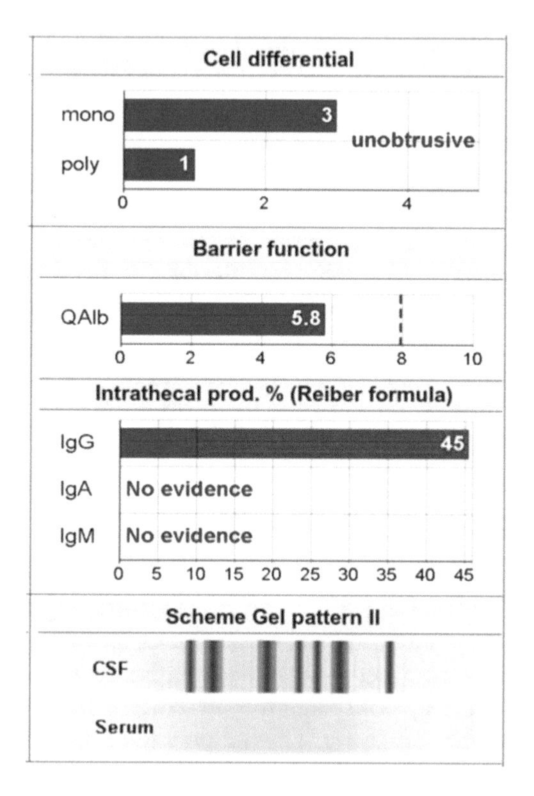

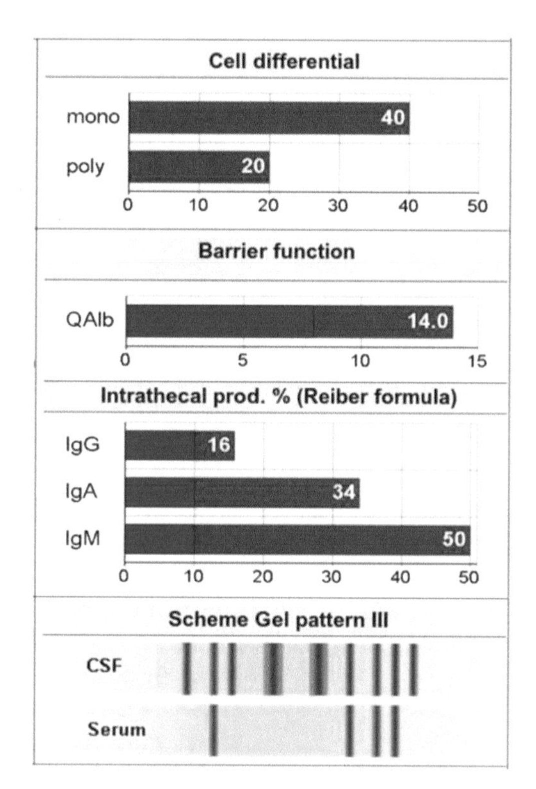

Fig. 9.1 Signature pattern of multiple sclerosis (left) versus neuroborreliosis (right)

and is the second most common clinical manifestation after erythema migrans which manifests in 95.4% [9]. Neurological involvement in Lyme disease must thus be recognized (Fig. 9.1b). The diagnosis is based primarily on clinical symptoms (radiculitis, meningitis, and neuritis, during the course complications like encephalitis or myelitis may develop) followed by a serological blood examination [10] and verified by CSF analysis. Patients with characteristic clinical symptoms that persist for more than 2–3 months without intrathecal antibody production are unlikely to suffer from neuroborreliosis. CSF cultivation of borrelia is difficult, depends upon the disease stage, and is only successful in 2–5% of patients. PCR has also only an inadequate sensitivity of 30–40% [11, 12]. Diagnosis is usually based on typical changes in the basic CSF analysis and verified with the antibody specificity index. Usually, a moderately increased cell count with a high percentage of B-lymphocytes and plasma cells is present. Often a three-class reaction (intrathecal IgG, IgA, and IgM synthesis) with IgM dominance can be found; blood–brain barrier dysfunction is typical (QAlb increased up to 50×10^{-3}) [13, 14]. These changes of barrier function and immunoglobulin synthesis present often even before any specific intrathecal borrelia antibodies is present and have a diagnostic sensitivity of 70% and specificity of 98% [13, 15]. The presence of acuity signs, i.e., an increased cell count, or a blood–brain barrier dysfunction combined with neurological symptoms indicate patient treatment. The recognition of these changes is facilitated by the visualization of the most relevant data. Basic CSF analysis presents with a characteristic signature pattern that needs to be complemented with the antibody specificity index to correctly diagnose the disease. A possible differential and in some cases additionally diagnosis is tick-borne encephalitis.

Treatment usually restores a normal cell count and normalizes blood–CSF barrier function. The dominant IgM three-class reaction, however, can persist as a "CSF scar" for decades [2, 16].

2 CSF Contaminated with Blood

The proteins and cells measured in CSF are present in much larger quantities as in blood. Any contamination of CSF with blood can be easily misinterpreted as intrathecal immunoglobulin synthesis (Fig. 9.2). The blood that may contaminate CSF origins from a vessel puncture during lumbar puncture. The peripheral blood cells and proteins increase artificially the cell number, the differential cell count, and the immunoglobulin concentrations. The CSF concentration of IgM is usually lower than the IgA-CSF concentration due to the its large molecular size, which is subsequently lower than the CSF-IgG concentration. This ratio changes with the artificial addition of blood and presents typically as a dominant but falsely elevated positive IgM synthesis. In extreme cases, the findings are very similar to the pattern seen in acute neuroborreliosis. Visible blood contamination of the CSF sample is the main indicator of falsely elevated intrathecal IgM synthesis. Blood contamination, however, is not always visible and the CSF sample may look inconspicuous. These samples often show no oligoclonal bands in isoelectric focusing despite an IgG (or multiclass) intrathecal production according to Reiber's formula or may present with a dominant IgM production lacking barrier dysfunction. The presence of blood can be easily checked with a dipstick (hemoglobin field) and the effect minimized with a mathematically correction using the exact number of erythrocytes and leucocytes present in blood and CSF [18].

3 Correction of CSF Leucocytes in Artificially Bloody CSF

Blood contaminated CSF can be corrected efficiently when the erythrocyte count is above 1000 cells/μL and below 7000–8000 cells/μL. The correction should also be applied for higher cell counts as well; in our opinion; but this delivers only a rough estimate of the actual CSF condition. Various correction procedures of diverse value and requirements exist. The best available method should always be used.

4 Correction of CSF Cells Contaminated with Peripheral Blood Cells

4.1 Simple Correction: Based on CSF Erythrocyte Count

In the simplest case, one leucocyte is subtracted for every 1000 erythrocytes present in the CSF. Mononuclear and polymorphic cell counts cannot be corrected (Table 9.1).

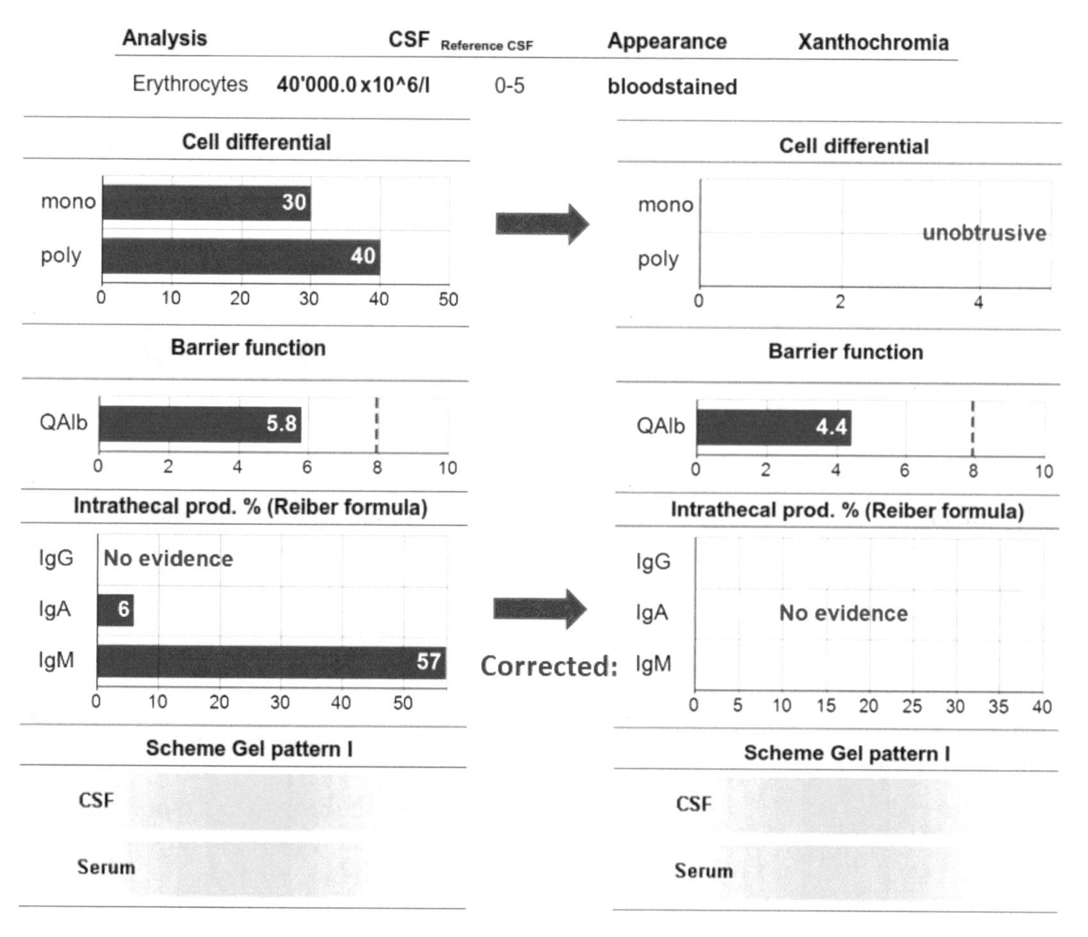

Fig. 9.2 False-positive intrathecal immunoglobulin caused by blood contamination. The correction removes the falsely elevated IgM concentration although the erythrocyte cell count exceeds 7000 cells largely

Table 9.1
Blood contamination: corrected CSF cell-count and -differential (patient example). The best correction method (3.) requires a full peripheral blood differential. Further explanation see text

			CSF leucocyte count correction:		
Cells/uL	**CSF**	**Blood**	**1. Simple**	**2. CBC Leuco**	**3. CBC Diff.**
RBC	40,000	4,200,00			
WBC	70	14,800	30	0	0
Polymorphic	40	10,750	?	?	0
Mononuclear	30	3800	?	?	0

- **Corrected Leucocytes** = Leuco CSF – (Ec CSF/1000)
  –> The number of leucocytes (WBC) decreases from 70 measured to 30 corrected cells (line 3, Table 9.1).

4.2 CBC Leuco: Leucocytes Corrected Using the Blood Differential

CSF white cell counts can be corrected using the blood differential of the patient. The sample for the blood differential should be obtained at a time point as closely as possible to the lumbar puncture. The difference should not exceed 24 h. Polymorphonuclear and mononuclear cell populations cannot be corrected. The leucocyte subpopulations are not available from the blood differential and only estimations are possible:

- **Corrected Leucocytes** = Leuco CSF – (Ec CSF * Leuco blood/Ec blood)
  –> CSF contains no leucocytes (WBC) after correction (line 2, Table 9.1).

4.3 CBC Diff. Leucocytes, Polymorphic and Mononuclear Cells Corrected Using the Blood Differential with WBC Differential Count

The best method uses the full blood differential, which contains and allows to correct polymorphic and mononuclear cell subpopulations. This type of correction is crucial in evaluating borderline white cell CSF elevations.

- **Corrected polymorphic cells** = Poly CSF – (Ec CSF × Poly Blood/Ec Blood)
- **Corrected mononuclear cells** = Mono CSF – (Ec CSF × Mono Blood/Ec Blood)

  Only this correction, based on the patient's blood differential including white cell differential, yields correct values when the CSF cell count is only slightly elevated. In the exemplary calculation in Table 9.1, only this method correctly classifies the cell count differential as reference. The elevated leucocytes, polymorphic or mononuclear cells in the CSF are caused by the pathological blood values and do not origin from the CSF.

  –> The CSF contains no leucocytes (WBC), polymorphic or mononuclear cells after correction (line 2–4, Table 9.1).

Table 9.2
Blood contaminated CSF: Effect of correction on protein concentration

CSF-protein	Measured mg/dL	Corrected mg/dL	Difference in %
Albumin	24.3	18.3	**–25**
IgG	4.6	2.88	**–37**
IgA	0.604	0.290	**–52**
IgM	0.167	0.054	**–68**

5 Correction of CSF Protein Concentration in Blood Contaminated Samples

CSF that is contaminated with blood can also artificially elevate Immunoglobulin CSF/serum quotients (Table 9.2) substantially. Oligoclonal bands in isoelectric focusing and the antibody index are rarely affected. This is most frequently observed with IgM due to its large molecular size (Fig. 9.2). Albumin and immunoglobulin concentrations can also be corrected:

- Corrected CSF protein = measured CSF protein – CSF Ec × protein blood/Ec blood).

  The calculated values must be considered an estimate and should be indicated as corrected. This is effective when the CSF erythrocyte count is between 1000 and 7000 cells/μL. Blood contaminated samples with more than 7000 erythrocytes/μL allow increasingly no reliable evaluation of albumin and immunoglobulin ratios and can still be misinterpreted as intrathecal synthesis [2, 17, 18].

6 Interpretation of the Antibody Index

Oligoclonal bands (OCB) occur frequently in viral and bacterial infections (neuroborreliosis OCB in 70%), autoimmune CNS disorders, or even psychiatric disease [19]. They are sensitive CSF markers of multiple sclerosis and occur in 90–98% of patients throughout the course of the disease. However, they are not specific for MS. Missing or occasionally transient OCB in the case of recurrent ocular and/or spinal cord involvement may indicate the presence of neuromyelitis optica (NMO) or NMO spectrum disease where OCB are present in around 30%. For differential

Table 9.3
Supplementary effect of methods over the time course of an HSV encephalitis

• Initial stage: PCR direct pathogen detection
– approximately first 9 days after initial symptoms
– sensitivity 75–98%, specificity 100%
• Late Stage: HSV antibody index
– approximately after 10 days of initial symptoms
– sensitivity 97%, specificity 73–100%
• Cell count 5–500 μL
– Initial stage: can present with mixed lymphocytic pleocytosis
– Reference cell count in 5%

diagnosis, OCB must thus be accompanied by other methods. A misdiagnosis of MS with associated therapy can, for instance, cause a substantial aggravation [20].

PCR allows for direct pathogen detection in the initial disease stages, whereas the diagnosis of later stage viral infections, approximately 10 days after initial symptoms, are better detectable by the specific quantitative antibody immune response (AI or ASI; "antigen-specific antibody index"). This can identify many diseases without direct pathogen detection, e.g., herpes virus or varicella zoster disorders, opportunistic infections, neuroborreliosis but also chronic inflammatory CNS conditions such as multiple sclerosis, optic neuritis, or autoimmune diseases with CNS involvement. Occasionally, the correct diagnosis of MS at initial clinical presentation can be difficult, particularly if another disease that resembles the symptoms of multiple sclerosis, for instance, acute disseminated encephalomyelitis (ADEM), might be present. The IgG-antibody-index (ASI) is part of the laboratory workup in these cases (Table 9.3). Simultaneous measurement of the antibody specificity indices against measles, rubella and varicella zoster virus ("MRZ"-reaction) and additionally herpes simplex allows to detect many acute or subacute infections. This can identify the characteristic polyclonal immune response of MS or the presence of a subacute varicella or herpes simplex infection.

The antibody specificity index is technically a demanding procedure. Measurement requires a high analytical sensitivity due to the low specific immunoglobulin concentration that is well below an already low total CSF-IgG concentration. ASI-values between 0.7 and 1.3 are considered reference range, ASI-values above 1.5 are usually considered pathological. Values below 0.6 are theoretically not very likely, but can occasionally occur with routine measurement. They are of no pathological significance, but the analysis should be reviewed for technical validity. Identical values due to antigen access can occur on the lower and upper range of the standard curve. ASI-values are usually reported numerically. This is not always possible when the CSF antibody concentration is below the analytical limit of detection. The ASI-index should than be clearly marked as "not computable." This documents to the clinician the correct measurement and avoids further inquiries.

Figure 9.3 shows some typical examples of accurate, reference range ASI-measurements. Values around the theoretical limit of 1.0 show valid measurement. Analytics should be reviewed (correct dilutions?) when results are consistently lower (Fig. 9.3a). ASI-results can also be uniformly distributed in the upper reference range, which is interpreted as absence of intrathecal production (Fig. 9.3b). However, the marked elevation of a single ASI- value above all other ASI-values that is still in the reference range must be interpreted as intrathecal synthesis (Fig. 9.3c).

Figure 9.4 shows the persistence of the MRZ reaction even 4 years after first diagnosis.

Reference range: Antibody Specifity Index 0.7–1.3: ASI-values in the reference range are close to the theoretical ASl value of 1.0 under analytically optimal conditions. The even distribution of ASI-values around the reference point confirms correct measurement under optimal conditions.

No intrathecal synthesis: The ASI of 1.4 is in line with the other ASI-values and is considered within the reference range.

Intrathecal synthesis: Although the ASI is in the reference range of (<1.5) this constellation is interpreted as the presence of intrathecal synthesis. The ASI of 1.3 for herpes simplex is much higher than the other measured ASI-values. An intrathecal immune response is present. This also illustrates that the solitary measurement of a single ASI is of limited diagnostic value.

Initial ASI-values

Same patient with persistent MRZ-reaction 4 years after initial diagnosis.

CSF/Serum IgG antibody index	Value	Reference	Chart
AI Herpes (IgG)	0.90	<1.5	0.9
AI Varizella (IgG)	0.90	<1.5	0.9
AI Rubella (IgG)	1.05	<1.5	1.1
AI Measles (IgG)	1.00	<1.5	1.0

Reference range: Antibody Index 0.7-1.3: ASI- values in the reference range are close to the theoretical AI value of 1.0 under analytically optimal conditions. The even distribution of ASI-values around the reference point confirms correct measurement under optimal conditions.

CSF/Serum IgG antibody index	Value	Reference	Chart
AI Herpes (IgG)	1.40	<1.5	1.4
AI Varizella (IgG)	1.20	<1.5	1.2
AI Rubella (IgG)	1.10	<1.5	1.1
AI Measles (IgG)	1.00	<1.5	1.0

No intrathecal synthesis: The ASI of 1.4 is in line with the other ASI-values and is considered within the reference range.

CSF/Serum IgG antibody index	Value	Reference	Chart
AI Toxoplasmosis	not computable	<1.5	
AI Epstein Barr	0.60	<1.5	0.6
AI Cytomegaly	0.70	<1.5	0.7
AI Herpes (IgG)	1.30 !	<1.5	1.3
AI Varizella (IgG)	0.60	<1.5	0.6
AI Measles (IgG)	0.70	<1.5	0.7

Intrathecal synthesis: Although the ASI is in the reference range of (<1.5) this constellation is interpreted as presence of intrathecal synthesis. The ASI of 1.3 for herpes simplex is much higher than the other measured ASI-values. An intrathecal immune response is present. This also illustrates that the solitary measurement of a single ASI is of limited diagnostic value.

Fig. 9.3 Interpretation of the antigen specificity index when values are in the reference range

CSF/Serum IgG antibody index	Value	Reference	Chart
AI Varizella (IgG)	0.80	<1.5	0.8
AI Rubella (IgG)	4.70	<1.5	4.7
AI Measles (IgG)	1.50	<1.5	1.5

Initial ASI-values

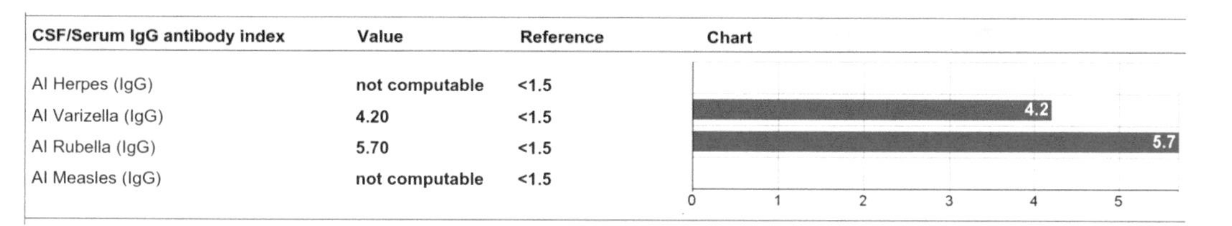

CSF/Serum IgG antibody index	Value	Reference	Chart
AI Herpes (IgG)	not computable	<1.5	
AI Varizella (IgG)	4.20	<1.5	4.2
AI Rubella (IgG)	5.70	<1.5	5.7
AI Measles (IgG)	not computable	<1.5	

Same patient with persistent MRZ-reaction 4 years after initial diagnosis

Fig. 9.4 Diagnostic significance of the ASI-index: This patient presented with minor CSF alterations (7 mononuclear cells, QALB slightly increased). Primarily, chronic neuroborreliosis was suspected. Based on the ASI final diagnosis consisted of chronic inflammatory autoimmune disease with a secondary elevation of ASI-levels

7 Biomarkers of Dementia

In most cases, basic CSF analysis does not show abnormalities when dementia is suspected. The pathophysiology of M. Alzheimer is characterized by the alteration of specific proteins, an increase of (p)tau and a decrease of amyloid beta 1-42 [21]. The related changes in tau and amyloid are difficult to recognize on a conventional report form due to their completely different reference ranges, the valuable addition of a fourth marker protein and the computation of the β-42/40 ratio [22, 23]. These markers of dementia form in many cases a characteristic pattern long before the development of dementia (Fig. 9.5), while the basic CSF marker profile detects or excludes other, potentially treatable diseases that can present clinically with a mentally confused state. In an earlier, own investigation with memory clinic outpatients who where assed for dementia ($n = 209$) we found in 37% results outside the reference range in basic CSF analysis. Forty-nine participants presented with a barrier dysfunction (QAlb $> 8, 23\%$), eight with an elevated cell count above 4 (4%), seven with dominant IgM synthesis, two with acute neuroborreliosis and one with acute tick-borne spring summer meningoencephalitis. Four patients had a

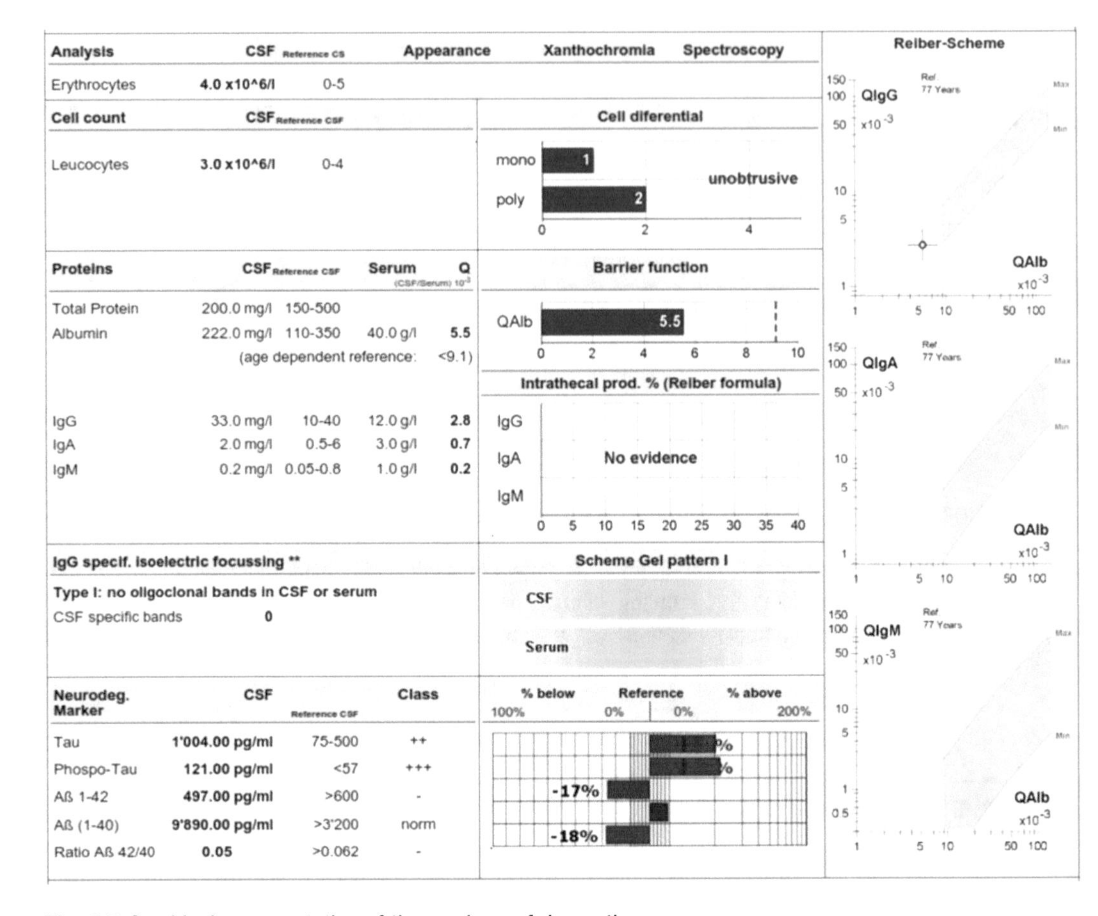

Fig. 9.5 Graphical representation of the markers of dementia

dominant IgA synthesis. The diagnosis was based on typical CSF findings in the basic marker profile and complemented by specific serology. The participants were all outpatients that received a checkup for dementia (Fig. 9.6).

8 Is CSF Analysis Worth the Effort?

CSF analysis is a complicated, complex, and cost-intensive procedure and criticized by some as an overuse of diagnostic procedures. Its value is best illustrated with a patient example (Fig. 9.7).

The 52-year-old female underwent lumbar puncture at her neurologist to verify a possible diagnosis of multiple sclerosis. Intrathecally produced immunoglobulins were only present in

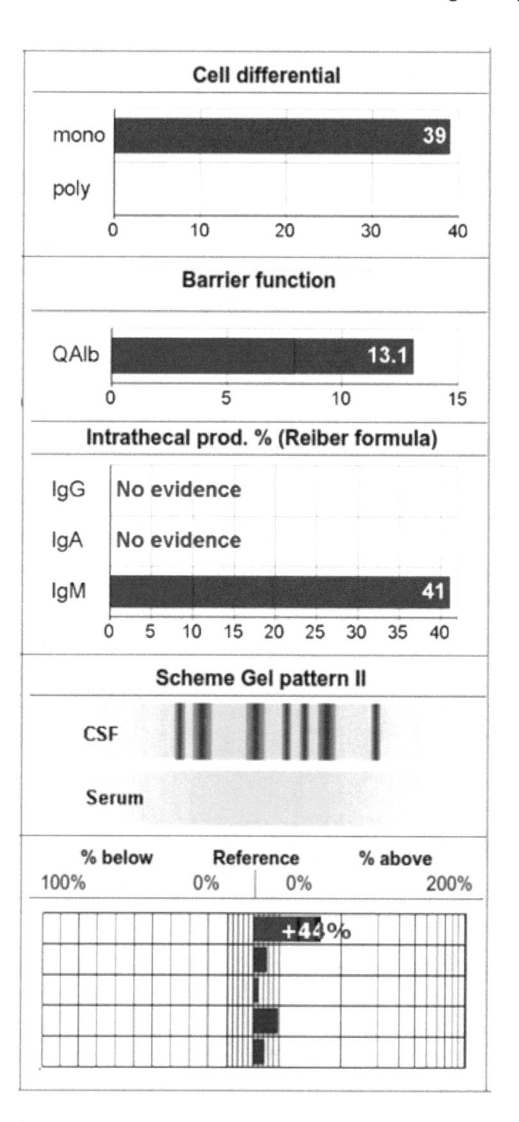

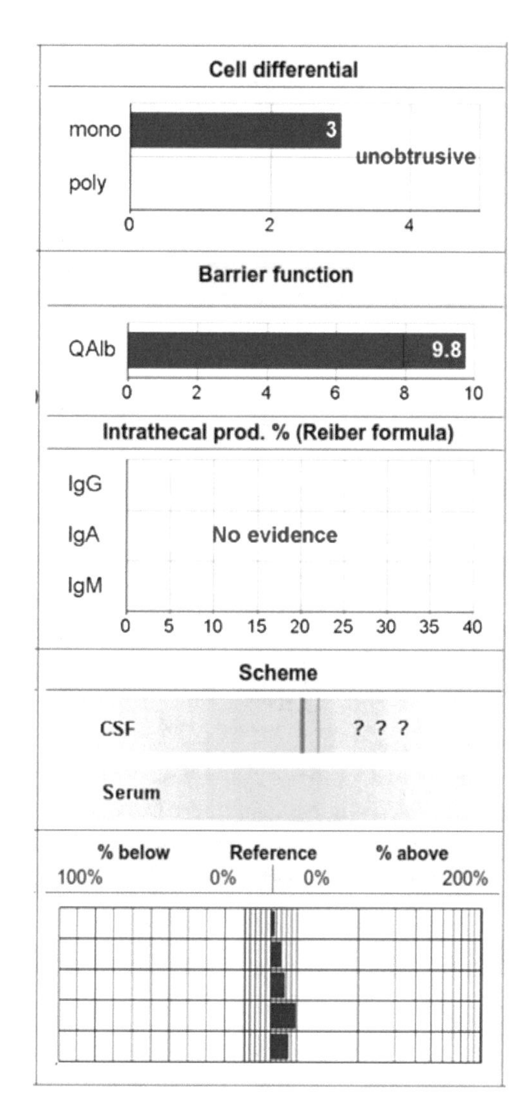

Fig. 9.6 Profile of an outpatient study participant that was checked for dementia but was diagnosed with acute tick-borne spring summer meningoencephalitis on initial presentation (left). CSF changes normalized 1 year later (right)

isoelectric focusing, which showed identical bands in serum and CSF and additionally oligoclonal bands only present in the CSF. Serum immunofixation confirmed the presence of a previously unknown faint monoclonal gammopathy (MGUS), type IgG-Lambda. The antibody specificity index for measles/rubella/varicella zoster (MRZ reaction) showed a polyclonal immune response. Final diagnosis consisted of multiple sclerosis combined with an unrelated and unknown monoclonal gammopathy of unknown significance. Thus, only the combination of multiple

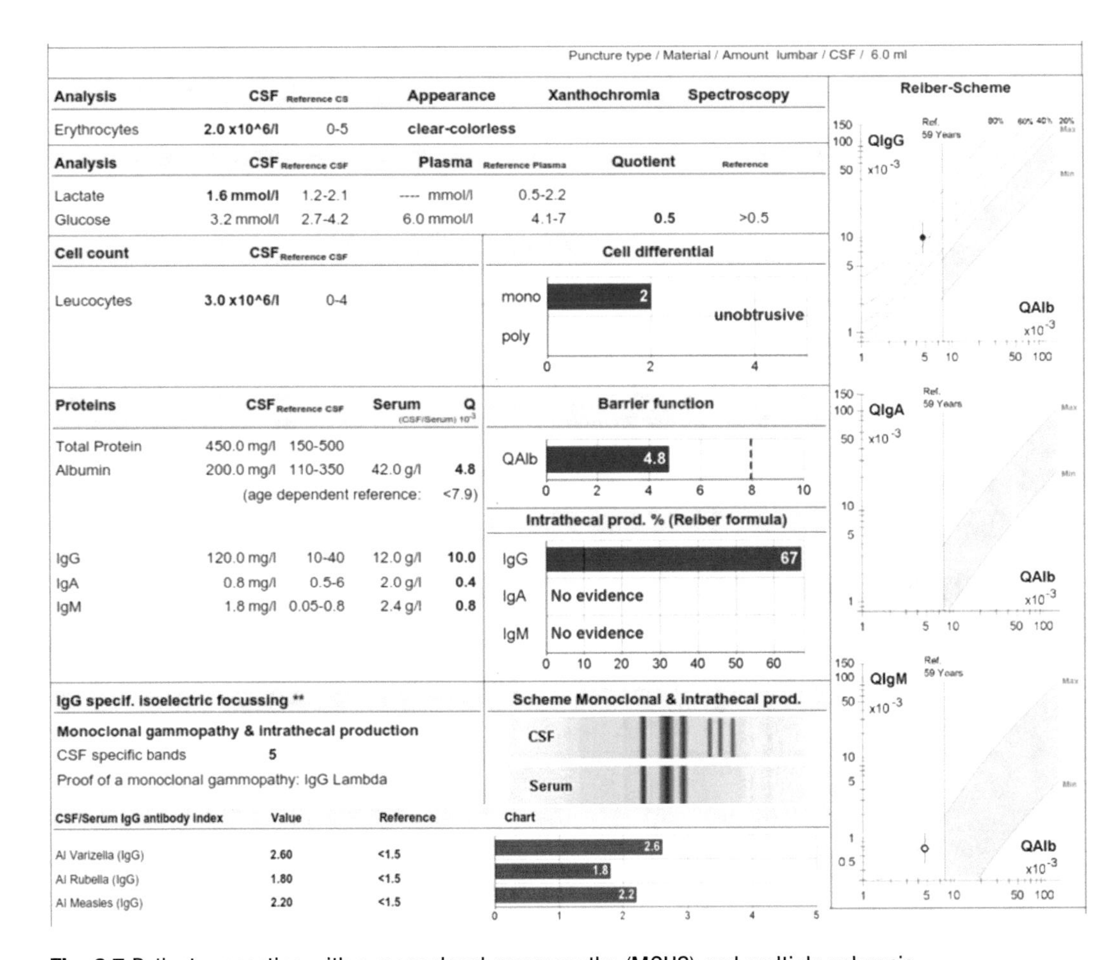

Fig. 9.7 Patient presenting with a monoclonal gammopathy (MGUS) and multiple sclerosis

results of serum and CSF analytes was able to establish diagnosis and initiate correct treatment.

9 Closing Statement

CSF analysis requires the measurements of numerous analytes in serum and CSF. The benefit for the patient is only present when the results are quickly conveyed, interpreted correctly by the physicians and adequate treatment initiated. The examples in this chapter are all courtesy of the CSF module of "MDI-Lablink"[1] [4, 5, 24].

The main goal of data visualization is to communicate information clearly and effectively [25]. Infographics in newspapers and smartphones transfer complex information clearly and efficiently in everyday life [26], while result communication is still a neglected

[1] http://mdilablink.com. Developed by the chapter authors, that gladly provide additional information.

area of laboratory medicine. The influence of graphical interpreted reports can probably be assessed best by a look at external orders. Reimbursed CSF analysis for other hospitals and neurologists in private practices increased 3 years after the introduction by 37% (2004: $n = 1021$; 2007: $n = 1402$) [5]. Ten years after its introduction external orders doubled ($n = 2104$) and made more than 75% of all orders for CSF analysis.

The introduction of such a separate system is difficult. Firstly, this is not a field of recognized research and only few systems are available. Secondly, it requires an investment where the benefit is not clear to hospital administrators. Today, most LIS system can also provide a Reiber's scheme, where the interpretation of intrathecal production and related analysis is in most cases left to the ordering clinician. CSF diagnostics is by some regarded as overuse of laboratory procedures. The value is scientifically proven numerous times, but when the results are inadequately presented in long rows of complex numbers is only decipherable by clinicians who are experts in the field. It is our firm belief that only a knowledge-based system combined with visual presentation of the relevant findings can unleash the true value of scientifically proven CSF analysis, where the textual information is validated by the graph and vice versa.

References

1. Reiber H (2016) Cerebrospinal fluid data compilation and knowledge-based interpretation of bacterial, viral, parasitic, oncological, chronic inflammatory and demyelinating diseases. Diagnostic patterns not to be missed in neurology and psychiatry. Arq Neuropsiquiatr 74 (4):337–350. issn:1678-4227
2. Reiber H, Peter JB Cerebrospinal fluid analysis: disease-related data patterns and evaluation programs. J Neurol Sci 184:101–122
3. Reiber H, Kruse-Sauter H, Quentin CD Antibody patterns vary arbitrarily between cerebrospinal fluid and aqueous humor of the individual multiple sclerosis patient: specificity-independent pathological B cell function. J Neuroimmunol 278:247–254. issn:1872-8421
4. Regeniter A et al (2003) Windows to the ward: graphically oriented report forms. Presentation of complex, interrelated laboratory data for electrophoresis/immunofixation, cerebrospinal fluid, and urinary protein profiles. Clin Chem 49:41–50
5. Regeniter A et al (2009) A modern approach to CSF analysis: pathophysiology, clinical application, proof of concept and laboratory reporting. Clin Neurol Neurosurg 111:313–318. issn:1872-6968
6. Andersson M et al (1994) Cerebrospinal fluid in the diagnosis of multiple sclerosis: a consensus report. J Neurol Neurosurg Psychiatry 57:897–902
7. Sindic CJ (1994) CSF analysis in multiple sclerosis. Acta Neurol Belg 94:103–111
8. Rauer S (2018) Leitlinien für Diagnostik und Therapie in der Neurologie. Stuttgart: Thieme, Vol. Vollständig überarbeitet: 21.
9. Wilking H, Stark K (2014) Trends in surveillance data of human Lyme borreliosis from six federal states in eastern Germany, 2009–2012. Ticks Tick-Borne Dis 5:219–224. issn:1877-959X
10. Koedel U, Hans-Walter P (2017) Lyme neuroborreliosis. Curr Opin Infect Dis 30:101–107. issn:1473-6527
11. Keller TL, Halperin JJ, Whitman M (1992) PCR detection of Borrelia burgdorferi DNA in cerebrospinal fluid of Lyme neuroborreliosis patients. Neurology 42:32–42
12. Picha D et al (2005) PCR in lyme neuroborreliosis: a prospective study. Acta Neurol Scand 112:287–292

13. Tumani H, Nolker G, Reiber H (1995) Relevance of cerebrospinal fluid variables for early diagnosis of neuroborreliosis. Neurology 45:1663–1670
14. Bednarova J (2006) Cerebrospinal-fluid profile in neuroborreliosis and its diagnostic significance. Folia Microbiol 51:599–603
15. Henkel K et al (2017) Infections in the differential diagnosis of Bellś palsy: a plea for performing CSF analysis. Infection 45:147–155. issn:1439-0973
16. Sindic CJ, Monteyne P, Laterre EC (1994) The intrathecal synthesis of virus-specific oligoclonal IgG in multiple sclerosis. J Neuroimmunol 54:75–80
17. Reiber H (2005) Proteindiagnostik. [ed.] Zettl. Klinische Liquordiagnostik, pp 177–200182
18. Reiber H (2011) Cerebrospinal fluid analysis report (Chapter 19). In: Wildemann B, Oschmann P, Reiber H (eds) Laboratory diagnosis in neurology, vol 24. Thieme Publishers, Stuttgart. issn:9783131441010
19. Endres D et al (2015) Immunological findings in psychotic syndromes: a tertiary care hospital's CSF sample of 180 patients. Front Human Neurosci 9:476. issn:1662-5161
20. Jarius S et al (2016) MOG-IgG in NMO and related disorders: a multicenter study of 50 patients. Part 2: Epidemiology, clinical presentation, radiological and laboratory features, treatment responses, and long-term outcome. J Neuroinflamm 13:280
21. Jack CR et al (2018) NIA-AA Research Framework: toward a biological definition of Alzheimer's disease. Alzheimers Dement 14:535–562. issn:1552-5279
22. Baldeiras I et al (2018) Addition of the Abeta42/40 ratio to the cerebrospinal fluid biomarker profile increases the predictive value for underlying Alzheimer's disease dementia in mild cognitive impairment. Alzheimers Res Ther 10:33
23. Lewczuk P et al (2017) Cerebrospinal fluid Abeta42/40 corresponds better than Abeta42 to amyloid PET in Alzheimer's disease. J Alzheimers Dis 55:813–822
24. Regeniter A et al (2009) Evaluation of proteinuria and GFR to diagnose and classify kidney disease: systematic review and proof of concept. Eur J Int Med 20:556–561. issn:1879-0828
25. Regeniter A et al (2000) Interpreting complex urinary patterns with MDI LABLINK: a statistical evaluation. Clin Chim Acta 297:261–273. issn:0009-8981
26. Aparicio M, Costa CJ (2015) Data visualization. Commun Des Q Rev 3:7–11. issn:2166-1200

Chapter 10

Method and Clinical Validation of Biomarkers for Neurodegenerative Diseases

Ulf Andreasson, Kaj Blennow, and Henrik Zetterberg

Abstract

In the Merriam-Webster dictionary, one definition of the word valid is "well-grounded or justifiable: being at once relevant and meaningful." Validation is then the process of determining the degree of validity. From this broad definition, it follows that validations can be made in many different fields with quite different implications. When talking about validation, it is therefore important to specify the subject under scrutiny and in this chapter the focus will be on validation of biomarkers.

Key words Biomarker, Neurodegeneration, Validation, Cut-point

1 Introduction

According to the National Institutes of Health working group on biomarkers definitions, a biomarker is "a characteristic that is objectively measured and evaluated as an indicator of normal biological processes, pathogenic processes, or pharmacologic responses to a therapeutic intervention" [1]. In the field of neurodegeneration, there are mainly two different types of biomarkers, fluid and imaging biomarkers, that are addressed. Fluid biomarkers can be any substance, e.g., a protein or a lipid, in any body fluid such as blood or cerebrospinal fluid. An image biomarker can be subdivided based on the technology used to create the image, e.g., positron emission or magnetic resonance. Irrespectively of the type of biomarker, two questions that need answers are: How reliable is the method and how can the result of the measurement be useful in a clinical or research setting? The first question is addressed in method validation while the other is the topic in clinical validation.

Charlotte E. Teunissen and Henrik Zetterberg (eds.), *Cerebrospinal Fluid Biomarkers*, Neuromethods, vol. 168,
https://doi.org/10.1007/978-1-0716-1319-1_10,

2 Method Validation

Even though the concept of method validation was established earlier, it was not until 1990 that a workshop was arranged on the topic for analytical methods. This workshop was cosponsored by five different organizations, Federation International Pharmaceutique, Health Protection Branch (Canada), American Association of Pharmaceutical Scientists, Association of Official Analytical Chemists, and the United States Food, and Drug Administration (FDA) and the scope was to provide guidance for validation of analytical methods [2]. Ten years later, the FDA published an official guidance built on this work [3]. The early work had its emphasis on drugs and their metabolites using liquid chromatography and in May 2018 the second version of the guidance was published, now also including sections on biomarkers and ligand-binding assays [4].

Medical laboratories can be accredited to ISO 15189:2012 [5], which states that method validation is the "Confirmation, through the provision of objective evidence, that the requirements for a specific intended use or application have been fulfilled." In this broad definition, there is no information on which experiments that need to be performed or how to evaluate them, which is likely due to the fact that different parameters need to be addressed depending on the method used. Some of the validation parameters, e.g., precision and measurement range, should however be addressed independently of the method under inspection.

In parallel with the work by FDA on method validation, there is a multitude of articles in scientific journals that address the topic, see for example references (6–12). Also the European counterpart to FDA, the European Medicines Agency, has published a guideline on bioanalytical method validation [13]. The Clinical and Laboratory Standards Institute (CLSI) has published many standards with instructions on how to evaluate different validation parameters [14]. Since CLSI is the appointed secretariat for the ISO technical committee, it might be that these standards will become international in the future. Sufficient to say is that at present the field of method validation is a living one and for all but a few there are different ways on how to evaluate the parameters.

Already during the method development the different parameters that are planned to be included in the validation should be kept in mind and tested. Not only to help in method optimization but also because before a validation is performed the acceptance criteria for each parameter has to be stated. Before a method validation is even considered there has to be an intended use or purpose for which the method can provide meaningful information. Therefore, some pilot studies should have been performed and the result should indicate that the method has usefulness.

There is one exception to this "rule" and that is when a method is used in a context that requires a validation by regulatory bodies, e.g., when the assay is employed for exploratory purposes in a clinical trial.

Listed below are some validation parameters with an explanatory text. There has been no attempt to exhaust all possible parameters, and the reader is referred to the literature cited above for details on how to perform needed experiments and how to present and interpret the results. However, for each parameter, we describe a simple experimental setup on how a relevant validation experiment could be performed.

2.1 Robustness

When evaluating the robustness (also called ruggedness) of a method small changes in the protocol for the method are investigated with respect to how they affect the end results. Factors to address are for example incubation times, temperature, pH, and concentration of reagents. There are usually many factors in a protocol that can be varied and a selection of the most crucial ones should be made to avoid unnecessary work. Also, to minimize the number of experiments, instead of testing one factor at the time, a factorial design can be utilized [15]. For commercial assays, this work should be done by the manufacturer and factors should be given with an interval, e.g., "Incubate for 30 ± 2 min," in the protocol.

2.2 Precision

Precision can be subdivided into three levels, repeatability, intermediate precision, and reproducibility. Repeatability (also called within-run or within-day precision) is the precision under minimal change in the conditions in a laboratory, i.e., having the same operator, batch of reagents, and instrument, in a short period of time while intermediate precision (also called between-run or between-day precision) is the precision when the conditions are altered. By definition, repeatability is always less than or equal to the intermediate precision. For the reproducibility, the laboratory is also a condition that is changed in addition to the ones for the intermediate precision.

Of the different validation parameters, precision is one of the few, the others being the related trueness and accuracy, that has been described in a standard from the international organization for standardization [16]. Even though the algorithm is based on analysis of variance (ANOVA), it is not completely straightforward to do the calculations. There are commercial softwares, e.g., Analyze-it® and StatisPro™, that can be used for the calculation of precision but a tool is also available free of charge [6].

There are different possible setups for the experiments that need to be performed. For example, five replicates on five different days or two replicates on 20 different days. It is important that the number of replicates is the same for all days in order for the

ANOVA to produce correct results. It is also important that the precision measurements are performed as if they are "real" samples, i.e., if the samples are to be run in duplicates then each replicate in the precision measurements has to be a duplicate as well.

2.3 Trueness

Trueness is the agreement between the average of an infinite number of measurements and that of a reference concentration. Since trueness is not a numerical quantity, it is expressed as bias, which is the systematic deviation from the expected value. For many analytes, there are no reference materials available, making the bias hard to estimate. In such instances, collaborative consortia may assemble, perform measurements on candidate reference materials using well-described methods (i.e., methods that are well documented but do not fulfill formal criteria for being accepted as certified reference methods) and assign the candidate reference material an analyte level (including variation) that may not be traceable back to a true concentration. This can be a tentative solution that allows for testing trueness also in the absence of a certified reference material. In principle the bias, if known, can be adjusted for and thereby eliminated.

The number of test results needed to estimate the uncertainty of the bias depend both on the predetermined magnitude of the bias and the repeatability for the method [17].

2.4 Selectivity

Selectivity is the ability of a method to correctly measure the analyte in a complex mixture. The interference of substances with similar physicochemical properties as the analyte can be tested by adding known concentrations of candidates to a sample and investigate if the addition has an effect on the measured concentration. It is difficult to be exhaustive with regard to all possible interfering substances since it might not be known a priori which these are. Another test for the selectivity is to spike known concentrations of the analyte to a sample and then investigate if the increase in measured concentration matches the added amount (the so-called spike-recovery).

2.5 Measurement Range

The measurement range is defined as the interval having the upper and lower limits of quantification (ULOQ and LLOQ) as endpoints. The LLOQ can be determined by running several blank samples, not containing any analyte, and calculate the mean value and standard deviation (SD) for the response signal. The LLOQ is then the measured concentration corresponding to the signal calculated from the mean value of the blank samples plus 10 SD. The measurement range can also be expressed as the interval of concentrations for which the precision lie below a certain limit. This limit should be connected to the precision of the method. Alternatively, the back-calculated concentrations from several calibrator curves can be used in the same way.

2.6 Dilution Linearity and Parallelism

Both dilution linearity and parallelism involve investigating the effect of dilution of a sample on the measured concentration of an analyte. The main difference between the parameters lies in that in dilution linearity the sample is spiked with a known amount of the analyte while no spiking is done in the parallelism experiment. The goal is to show that the analyte in a sample with a concentration above the ULOQ can be reliably quantitated after dilution. Since the composition of the diluent, used for preparing the calibrator curve, might be quite different from the one in the sample this is not necessarily true due to what is called the matrix effect. For ligand-binding assays, the signal can be saturated or even decreased (Hook effect) at high concentrations, and this behavior can be evident after a dilution linearity experiment.

2.7 Stability

Pre-analytical factors have the potential to affect the results in a measurement to a high degree, and it might be a need for standardization with regard to sample procedure and treatment. To test the latter fresh samples should be subjected to different storage conditions and freeze/thaw cycles to evaluate the best handling. The different storage conditions should mimic possible scenarios that can happen in a laboratory such as hours at room temperature, days in the fridge, and months in a freezer. The results can serve as a basis for the shipment instructions and proper handling in the laboratory.

2.8 Quality Controls

To establish an internal quality control (QC) scheme is strictly speaking not a validation parameter. However, a validation is performed during a limited time frame and is therefore a kind of snapshot of the performance. To be able to monitor the performance of the assay over time, especially with regard to the precision, it is advisable to have QC samples prepared to be used when the assay is taken into service. Most convenient would be to have aliquots of the same samples used for the evaluation of the precision since the expected value is then already established. Otherwise, this has to be done before unrelated QC samples are introduced. A common way to approve a run based on the QC samples is to use the framework of the Westgard rules [18].

2.9 Validation Report

The results of a validation need to be summarized in a validation report in which the acceptance criteria for the different parameters should be stated prior to the execution of the experiments. A template for the validation report can be found in reference (6). Together with the validation report, it is important to file also the protocol for the method and the source data. The protocol shows which version of the method that was validated and if the method is changed in any way there is a need to consider a new validation whose extent depends on the changes made. The source data, usually printouts from the instrument software used for data acquisition and analysis, need to be signed and dated. These three parts constitute the documentation of the method validation and should be bundled together for future reference.

3 Clinical Validation

Clinical validation is the process of determining the degree of validity of a biomarker in a clinical or research setting, i.e., in a context which is relevant to the intended use or purpose of the biomarker. As is the case for method validation, a number of key parameters should be considered. These are detailed below. When interpreting the results for each parameter, it is important to consider the context of use of the biomarker. As an example, a screening test for a disease that may be fatal if left untreated should have high sensitivity but the specificity may be of less importance, at least if the diagnosis can be substantiated using more specific tests in a sequential testing algorithm. In explorative biomarker discovery work, the context of use of a biomarker candidate may be unknown. However, it may emerge in studies testing the validation parameters discussed below.

3.1 Biological Variation

Early in the development process, a biomarker candidate should be examined for biological variation. What is the diurnal variation of the biomarker and what is its stability over days, weeks, months, or years? What is a clinically meaningful change in a biomarker concentration? What factors, not related to the process that the biomarker is thought to reflect, may influence its concentration? Regarding inter-individual variation, potential gender and age effects have to be examined. However, it is hard to produce a generic list of what other intra- and inter-individual biological factors that should be examined; this will largely depend on the type and context of use of the biomarker.

3.2 Sensitivity

Sensitivity is defined as the proportion of subjects with the disease and a positive test result in relation to the total group of subjects with the disease (Table 10.1). It is determined using samples from patients who are known to have the disease or the pathological process that the biomarker is thought to reflect. The diagnosis has to be made using an accepted reference standard. Thus, sensitivity does not depend on disease prevalence; a non-disease control group is not needed. In research on neurodegenerative diseases, a neuropathological diagnosis is often considered the gold standard. For autosomal dominant diseases, a genetic test may suffice. However, as a biomarker may reflect onset of a pathological process that may not exist at birth in mutation carriers, a reference standard for the pathology that the biomarker is thought to reflect may still be important. In research on Alzheimer's disease, amyloid positron

Table 10.1
Numbers of subjects classified according to disease status and dichotomized using a biomarker test

	Biomarker-positive	Biomarker-negative
Subjects with the disease	True positive (TP)	False negative (FN)
Subjects without the disease	False positive (FP)	True negative (TN)

Sensitivity = TP/(TP + FN)
Specificity = TN/(TN + FP)
Positive predictive value = TP/(TP + FP)
Negative predictive value = TN/(TN + FN)
Accuracy = (TP + TN)/(TP + TN + FP + FN)

emission tomography of amyloid β (Aβ), validated against neuropathology, has made it easier to define sensitivity (and specificity, see below) for Aβ-related biomarkers in body fluids [19]. In many research settings, the sensitivity of a biomarker test is defined using patients who have received a clinical diagnosis of the disease. Ideally, the diagnosis should have been made using internationally accepted consensus criteria. Preferably, these should have been validated against a neuropathological diagnosis but in practice, they are never 100% accurate; a biomarker can never be better than the reference standard used to make the diagnosis. Notably, the reference standard, whatever it is, must not include any information related to the biomarker to be examined (in clinical validation literature, the biomarker is often called "the index test").

3.3 Specificity

Specificity is defined as the proportion of subjects without the disease with a negative test result in relation to the total number of subjects without disease (Table 10.1). A numerical value on specificity may be determined using samples from subjects who do not have the disease or pathological process that the biomarker is thought to reflect. The population may be age- and gender-matched healthy controls but may also be subjects suffering from other conditions that do not involve pathophysiological changes related to the biomarker. No patients with the condition are needed. As with sensitivity, the reference standard used to diagnose the test population is important; to define the population optimal for testing specificity, the reference standard should effectively exclude the disease or pathological process that the biomarker is thought to reflect.

3.4 Accuracy

The accuracy of a test is the proportion of subjects who are correctly classified, i.e., the sum of true positives and true negatives divided by the total number of tested subjects (Table 10.1). As with sensitivity and specificity, diagnostic accuracy is not affected by disease prevalence.

3.5 Predictive Values

Positive and negative predictive values quantify the probability of disease given a subject's test result (Table 10.1). These quantities depend on the disease prevalence in the target population. Predictive values are thus relevant only if the proportion of disease subjects in the sample is representative of disease prevalence in the target population. A biomarker test examined in a population with very high prevalence of the disease will always get a high positive predictive value, while a biomarker tested in a population with a very low disease prevalence will always get a very high negative predictive value. Ideally, predictive values should be tested in a setting that resembles the clinical or research setting in which the biomarker is intended to be used as closely as possible. Data from case-control studies may be used to simulate predictive values at different disease prevalences, but these should be validated in real-life settings.

3.6 The Receiver Operating Characteristic (ROC) Curve

One way of visualizing the diagnostic performance of a biomarker is the ROC curve that plots sensitivity against 1-specificity over the range of possible cut-points (Fig. 10.1). ROC curves are often used to show a biomarker's value for detecting a binary outcome (e.g., case/control status) and are sometimes summarized by the area under the ROC curve. The latter has an appealing simple interpretation: the higher the area the better the diagnostic accuracy of the biomarker.

3.7 Likelihood Ratios

The positive (or negative) likelihood ratio is the ratio of the probability of a positive (or negative) test result in subjects with the disease to the probability of a positive (or negative) test in non-disease subjects.

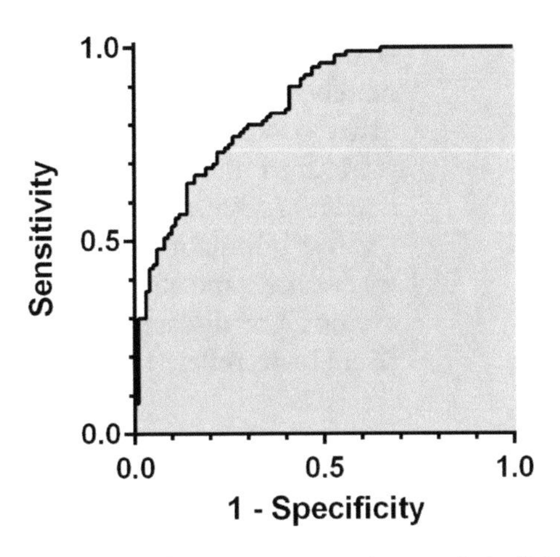

Fig. 10.1 An example of a receiver operating characteristic (ROC) curve where the area under the curve is shaded

The positive likelihood ratio (LR+) can be simply calculated according to the following formula: LR+ = sensitivity/(1−specificity), while the negative likelihood ratio (LR−) is = (1−sensitivity)/specificity.

Like sensitivity and specificity, likelihood ratios do not depend on disease prevalence. A useful property of likelihood ratios is that the odds of a subject having the disease given a positive test result are simply the pre-test odds (prevalence/(1−prevalence)) of disease multiplied by the positive likelihood ratio. Likelihood ratios can thus be used to calculate the odds (and thus probability) of disease given the test result in populations with differing disease prevalence. Likelihood ratios also make it possible to use probabilistic reasoning in clinical practice, i.e., when information from several tests are used to make clinical decisions. This is also called Bayesian reasoning, being based on Bayes' theorem, in which the probability of a hypothesis is modified by further data.

3.8 Cut-Point Determination

New proposed criteria for the diagnosis of Alzheimer's disease, and increasingly other neurodegenerative diseases, incorporate biomarkers, most of which are normally measured on a continuous scale. Operationalizing such criteria thus requires continuous biomarkers to be dichotomized, which in turn requires selection of a cut-point at which to dichotomize. As will become clear below, cut-point determination is a science in itself.

In clinical chemistry, a common way of determining if a biomarker test is positive or negative is to define reference limits. Here, the biomarker is measured in people who are healthy or at least do not have the disease or pathological process the biomarker is thought to reflect. Using the data, a 95% confidence interval of the biomarker results may be created. If a patient has a value outside this range, the test is "positive." Clearly, this does not necessarily mean that the patient has the condition but together with additional information, a diagnosis may be made. For cut-point determinations, the intended use of the biomarker is important. A test designed to discover a treatable disease that is fatal if missed should favor sensitivity over specificity, and vice versa if making the diagnosis could lead to potentially harmful interventions.

A number of additional methods have been suggested to facilitate cut-point selection. One is Youden index, which is defined as the sum of sensitivity and specificity minus one. The cut-point resulting in the highest Youden index, would be the preferred one. However, again this may be an oversimplification, as cut-point selection depends on the context of use. Another approach sometimes used is to find the cut-point whose point on the ROC curve lies closest to the top-left corner (0,1). The rationale for this is that the ROC curve for a perfect biomarker extends to the top-left corner, and this method thus determines the cut-point that is closest to "perfection." However, similar to Youden index, this approach is equivalent to maximizing a function of sensitivity and specificity.

Various attempts are being made to get away from dichotomizing cut-points. The approach closest to clinical implementation is the definition of gray zone values or confidence intervals around cut-points. Such gray zones, where it actually is impossible to determine if a test is positive or negative, are influenced by both analytical and biological variation and are something many clinicians develop an intuitive feel for with experience. Another clinical action in regard to diagnostic uncertainty is to resample the patient and look for disease-related changes. This is often not formalized and requires experience and good knowledge on the analytical and within-subject biological variation of biomarker concentrations.

3.9 Clinical Robustness

When a biomarker has been technically and clinically validated in one laboratory, as well as in inter-laboratory settings and independent, clinically relevant cohorts, it is important to test the biomarker in clinical practice. Biomarkers that have been newly introduced in clinical laboratory practice have to be carefully monitored for longitudinal stability using internal and external QC samples and interpreted humbly to detect overseen confounders that may emerge when individuals from the general population or from patients with diseases not previously examined receive the test. This type work is essential to establish the clinical robustness of the biomarker. It may also give exciting new research ideas and increase our understanding of what the biomarker actually represents. One concrete example of this is how we no longer consider cerebrospinal fluid tau a general biomarker for neurodegeneration but rather a relatively Alzheimer-specific biomarker for altered tau metabolism and secretion from neurons that are still alive. The first results indicating this emerged in studies showing negative results for CSF tau in non-Alzheimer's neurodegenerative diseases [20]. The revised interpretation was finally corroborated in a convincing manner by direct assessment of tau turnover in humans with and without Aβ pathology [21].

4 Closing Remark

Biomarkers are playing an increasingly important role in the diagnosis and monitoring of Alzheimer's disease and other neurodegenerative disease in clinical practice, therapeutic trials, and observational studies. It is very hard to interpret biomarker results without knowledge on assay characteristics and clinical performance in settings relevant the context in which the biomarker is being used. In this chapter, we have discussed parameters of importance in analytical and clinical validation of biomarkers. We have also discussed how to design studies to define and test such parameters. Biomarker validation is complex and requires profound knowledge on exactly what the biomarker is thought to represent. It is an iterative process and it is hard to determine what is needed before a biomarker may be labelled fully validated. Once a biomarker test has been determined reasonably fit for purpose, test performance has to be continuously monitored and revalidation experiments may have to be initiated as new data come in.

References

1. Biomarkers Definitions Working Group (2001) Biomarkers and surrogate endpoints: preferred definitions and conceptual framework. Clin Pharmacol Ther 69(3):89–95
2. Shah VP, Midha KK, Dighe S, McGilveray IJ, Skelly JP, Yacobi A et al (1992) Analytical methods validation: bioavailabilitybioequivalence and pharmacokinetic studies pharmaceutical research. J Pharm Sci 9(4):588–592
3. Guidance for Industry (2001) Bioanalytical method validation. Fed Reg 66:28526–28527
4. Guidance for Industry (2018) Bioanalytical method validation. Fed Reg 83:23690–23691
5. ISO 15189:2012 (2012) Medical laboratories—Requirements for quality and competence
6. Andreasson U, Perret-Liaudet A, van Waalwijk van Doorn L, Blennow K, Chiasserini D, Engelborghs S et al (2015) A practical guide to immunoassay method validation. Front Neurol 6:179
7. Chau CH, Rixe O, McLeod H, Figg WD (2008) Validation of analytic methods for biomarkers used in drug development. Clin Cancer Res 14(19):5967–5976
8. Cullen VC, Fredenburg RA, Evans C, Conliffe PR, Solomon ME (2012) Development and advanced validation of an optimized method for the quantitation of Abeta42 in human cerebrospinal fluid. AAPS J 14(3):510–518
9. DeSilva B, Smith W, Weiner R, Kelley M, Smolec J, Lee B et al (2003) Recommendations for the bioanalytical method validation of ligand-binding assays to support pharmacokinetic assessments of macromolecules. Pharm Res 20(11):1885–1900
10. Findlay JW, Smith WC, Lee JW, Nordblom GD, Das I, DeSilva BS et al (2000) Validation of immunoassays for bioanalysis: a pharmaceutical industry perspective. J Pharm Biomed Anal 21(6):1249–1273
11. Khan MU, Bowsher RR, Cameron M, Devanarayan V, Keller S, King L et al (2015) Recommendations for adaptation and validation of commercial kits for biomarker quantification in drug development. Bioanalysis 7(2):229–242
12. Lee JW, Devanarayan V, Barrett YC, Weiner R, Allinson J, Fountain S et al (2006) Fit-for-purpose method development and validation for successful biomarker measurement. Pharm Res 23(2):312–328
13. EMA (2011) Guideline on bioanalytical method validation
14. CLSI (2018) [Collection of standards]. www.clsi.org
15. Sittampalam GS, Smith WC, Miyakawa TW, Smith DR, McMorris C (1996) Application of experimental design techniques to optimize a competitive ELISA. J Immunol Methods 190 (2):151–161
16. ISO 5725-2 (1994) Accuracy (trueness and precision) of measurement methods and results—Part 2: basic method for the determination of repeatability and reproducibility of a standard measurement method
17. ISO 5725-4 (1994) Accuracy (trueness and precision) of measurement methods and results—Part 4: basic methods for the determination of the trueness of a standard measurement method
18. Westgard JO, Barry PL, Hunt MR, Groth T (1981) A multi-rule Shewhart chart for quality control in clinical chemistry. Clin Chem 27 (3):493–501
19. Blennow K, Mattsson N, Scholl M, Hansson O, Zetterberg H (2015) Amyloid biomarkers in Alzheimer's disease. Trends Pharmacol Sci 36(5):297–309
20. Itoh N, Arai H, Urakami K, Ishiguro K, Ohno H, Hampel H et al (2001) Large-scale, multicenter study of cerebrospinal fluid tau protein phosphorylated at serine 199 for the antemortem diagnosis of Alzheimer's disease. Ann Neurol 50(2):150–156
21. Sato C, Barthelemy NR, Mawuenyega KG, Patterson BW, Gordon BA, Jockel-Balsarotti J et al (2018) Tau kinetics in neurons and the human central nervous system. Neuron 98 (4):861–864

Chapter 11

Sample Preparation for Proteomic Analysis of Cerebrospinal Fluid

Johan Gobom

Abstract

This protocol describes sample preparation of CSF for proteomic analysis by LC-MS. The protocol can optionally be used in conjunction with quantification by the tandem mass tag (TMT) technique, which due to its high multiplexing capability is particularly useful in discovery proteomic studies when large cohorts are to be analyzed.

Key words Proteomics, Protein identification, Protein quantification, Biomarker discovery, LC-MS

1 Introduction

CSF samples are treated with trypsin to produce proteolytic peptides that can be analyzed by LC-MS for peptide- and protein identification and quantification. In the first step, the detergent sodium deoxycholate is added to the CSF to partially denature proteins and break up protein interactions, thereby increasing the proteins' accessibility to proteolytic cleavage. Solubilizing the CSF proteins is a critical step in the analysis; due to their different physico-chemical properties, different proteins require different detergents and chaotropes to facilitate their solubilization. While there is not one single solubilizing agent that enables analysis of all CSF proteins, deoxycholate has been found to yield reproducible proteolysis and overall large number of proteins detected by LC-MS, and has been used in several recent CSF proteomic studies [1–3]. Protein disulfides are reduced by incubation with tris(2-carboxyethyl)phosphine (TCEP), and cysteine residues are subsequently carbamidomethylated using iodoacetamide to prevent re-forming of disulfide bonds. Following tryptic digestion, the samples are acidified to precipitate the deoxycholate, which is then removed by centrifugation, and the supernatants are desalted using C_{18} solid-phase extraction (SPE). Sample aliquots can be

Charlotte E. Teunissen and Henrik Zetterberg (eds.), *Cerebrospinal Fluid Biomarkers*, Neuromethods, vol. 168,
https://doi.org/10.1007/978-1-0716-1319-1_11,

analyzed directly by LC-MS but due to the high sample complexity and wide range of concentrations of proteins in CSF, significantly more peptides are identified and accuracy of quantification is improved if a peptide fractionation step is included prior to LC-MS. Here, we describe a fractionation method based on high-pH reversed phase (HpH-RP) chromatography with fraction concatenation [4, 5]. The protocol is compatible with multiplex isobaric labeling using the tandem mass tag (TMT) method. TMT labeling has been included in the protocol as an optional step after tryptic digestion.

2 Materials

Use reagents of high analytical grade and ultra-high purity water for all solutions. Handle CSF and reagents in polypropylene tubes. Use "protein-low-binding" polypropylene tubes to handle SPE purified samples and HPLC fractionated peptides.

1. Sample denaturation, reduction, and alkylation
   (a) DOC/TEAB: 5% (w/v) Sodium deoxycholate (DOC), 0.5 M tri-ethyl ammonium bicarbonate (TEAB). Stable at 4 °C for at least 2 weeks.
   (b) TCEP: 25 mM tris(2-carboxyethyl)phosphine (TCEP) in DOC/TEAB. Prepare fresh.
   (c) Iodoacetamide solution: 400 mM iodoacetamide, 100 mM TEAB. Light sensitive; prepare fresh.
2. Tryptic digestion
   (a) trypsin (proteomic grade)
3. TMT labeling (optional, for quantification)
   (a) TMT labeling reagents
   (b) acetonitrile
   (c) 5% Hydroxylamine: Dilute 50% hydroxylamine to 5% with 100 mM TEAB. This solution is stable at 4 °C for at least 2 weeks.
4. Removal of deoxycholate and dilution
   (a) hydrochloric acid (0.5 M)
5. Desalting of tryptic peptide samples
   (a) Vacuum manifold for SPE.
   (b) Sep-Pak C18 cartridge (Waters, 50 mg sorbent)
   (c) SPE Buffer A: 0.1% tri-fluoroacetic acid. Stable at 4 °C for at least 2 weeks.
   (d) SPE Buffer B: 80% acetonitrile, 0.1% tri-fluoroacetic acid. Stable at 4 °C for at least 2 weeks.

6. Offline high-pH reverse phase HPLC sample fractionation
   (a) HPLC system equipped with a ternary gradient pump operated at 100 μL/min, a fraction collector operating on time-based fractionation, and a UV detector set to measure absorbance at 214 nm.
   (b) Chromatographic separation column: XBridge column (Waters, BEH130 C18 3.5 μm, 2.1 mm × 250 mm), which is stable at basic pH.
   (c) HpH-RP Buffer A: water
   (d) HpH Buffer B: 84% acetonitrile (v/v)
   (e) HpH Buffer C: 25 mM NH_4OH
   (f) HpH Loading buffer: 2.5 mM NH_4OH, 2% acetonitrile (v/v).

3 Methods

Sample Denaturation, Reduction, and Alkylation

1. Thaw CSF by gentle agitation at RT (*see* **Note 1**).
2. Take out 50 μL CSF for proteomic analysis (*see* **Note 2**).
3. To each CSF sample (50 μL), add 13 μL TCEP solution.
4. Vortex the samples for 10 min at RT.
5. Incubate the samples for 1 h at 55 °C. While the samples are incubating, prepare the iodoacetamide solution.
6. Allow the samples to cool to RT.
7. Add 1.6 μL iodoacetamide solution to the samples, vortex briefly.
8. Incubate the samples for 30 min at RT in the dark.

Tryptic Digestion

1. Dissolve trypsin in the resuspension buffer provided by the manufacturer to a concentration of 0.2 μg/μL (*see* **Note 3**).
2. Add 13 μL trypsin solution to each sample.
3. Incubate the samples over night at 37 °C.

TMT Labeling (Optional)

1. Take out the TMT labeling reagents of the freezer allow them to reach room temperature before opening the vials (*see* **Notes 4** and **5**).

2. Dissolve the TMT labeling reagents in acetonitrile to a concentration of 19.5 μg/μL, vortex and spin down the liquid in a table centrifuge.
3. Add 7.5 μL of the TMT reagent solution to each sample, vortex briefly.
4. Incubate the samples for 1 h at RT on a shaker.
5. Add 6 μL 5% hydroxylamine solution to quench the labeling process, vortex briefly.
6. Incubate the samples for 30 min RT on a shaker.
7. Combine the samples into TMT multiplex sets.

Removal of Deoxycholate and Dilution

1. Add a volume, corresponding to 1/10 of the sample volume, of 0.5 M HCl to acidify the samples. This causes DOC to precipitate.
2. Add a volume of 0.1% TFA to the samples, sufficient to decrease the acetonitrile concentration in the samples to <3%.
3. Centrifuge at 30,000 ×*g* for 15 min at 4 °C. Pipet off the supernatant and transfer it to another tube.

Desalting of Tryptic Peptide Samples

This step is performed by solid phase extraction (SPE) using reversed-phase C_{18} cartridges operated using a vacuum manifold. Follow the manufacturer's instructions regarding general operation.

1. Wash the SPE cartridges with 2*1000 μL SPE Buffer B.
2. Equilibrate the SPE cartridges with 2*1000 μL SPE Buffer A.
3. Load the samples.
4. Wash the SPE cartridges with 2*1000 μL SPE Buffer A.
5. Place empty vials under SPE cartridges.
6. Elute with 1000 μL SPE Buffer B.
7. Lyophilize the eluates by vacuum centrifugation.

Offline High-pH Reverse Phase HPLC Sample Fractionation

1. Dissolve the peptide samples in 16 μL (or other suitable injection volume for the used HPLC system) HpH-RP Loading buffer by shaking for 20 min.
2. Load the samples on the HPLC and perform peptide separation by applying the gradient shown in Table 11.1.

Table 11.1
LC-gradient used for high-pH reversed-phase fractionation

t (min)	%A	%B	%C	Comment
0	89	1	10	Injection
4	89	1	10	Start fraction collection
65	45	45	10	Stop fraction collection
66	0	90	10	Clean the column
76	0	90	10	Clean the column
76.1	89	1	10	Equilibrate the column
86	89	1	10	End

3. Collect 1-min fractions. When fractionation is complete, concatenate fractions by combining every 24th fraction (i.e., Fraction 1 + 25 + 49, 2 + 26 + 50 etc.). In the system used in the author's laboratory, this was achieved by circling over two rows in a 96-well microtiter plate. The resulting 24 concatenated fractions contain peptides from three fractions separated by 24 min retention time (*see* **Note 6**).
4. Lyophilize the eluates by vacuum centrifugation at and store the samples at −80 °C prior to LC-MS analysis.

4 Notes

Notes relating to the overall protocol:

Proteomic analysis by LC-MS is very sensitive to sample contamination. A frequently occurring class of contaminants are polyethylene glycol (PEG)-based polymers. Small amounts present in the samples give rise to series of intense MS peaks distributed over the chromatographic separation, characterized by mass differences of 45 Da between peaks, corresponding to the PEG monomer (O–CH2–CH2). The polymers hamper the detection of sample peptides by using up the binding capacity of the nano-LC column and causing ion suppression. PEG-based polymers can come from a wide variety of sources, including plastic softeners, rubber gloves, and skin lotion. If encountered, test each step of the sample preparation procedure separately using blank samples and replace contaminated solutions, reagents, or plasticware. Keratins, e.g., from hair, skin, dust, and wool sweaters, are another frequently encountered source of contamination. While keratin-derived peptides can be sorted out in data processing, they may still impair the LC-MS analysis. Thus, it is important that the work place and laboratory equipment is kept clean and dust-free.

Notes specifically relating to the "Materials" and "Methods" section:

1. CSF samples should be visually inspected for discoloration that may indicate blood contamination resulting from puncture bleeding. The significantly higher protein concentration in blood and the presence of proteases may greatly affect the analytical results.
2. In order to compare quantitative results across multiple TMT multiplex sets, a reference sample should be included in one TMT channel in all sets. The reference sample is generally produced by pooling an aliquot of all study samples. For the reference sample, take out from each study sample 50 μL * number of TMT sets * 1.2 (for pipetting overhead).
3. While trypsin is the most widely applicable protease, it is not suitable for all proteins; in some proteins or regions of proteins, Lys and Arg residues may be too closely or too sparsely spaced to produce cleavage products of suitable length for mass spectrometric identification. In such cases, alternative proteases such as Asp-N or Glu-C may be used instead or as complement.
4. Opening the TMT reagent vials while they are still cold may cause humidity from the air to precipitate in the reagent vials, which may inactivate the reagent.
5. Take care to note the lot number of the TMT reagents as this information is necessary to correct for isotope impurities when quantifying peptides based on TMT reporter ion signals.
6. Fraction concatenation reduces the number of samples to be analyzed by LC-MS while ensuring that each LC-MS sample contain peptides evenly distributed over the entire chromatographic separation to optimize mass spectrometric data acquisition.

References

1. Magdalinou NK, Noyce AJ, Pinto R, Lindstrom E, Holmen-Larsson J, Holtta M, Blennow K, Morris HR, Skillback T, Warner TT, Lees AJ, Pike I, Ward M, Zetterberg H, Gobom J (2017) Identification of candidate cerebrospinal fluid biomarkers in parkinsonism using quantitative proteomics. Parkinsonism Relat Disord 37:65–71
2. Pavelek Z, Vysata O, Tambor V, Pimkova K, Vu DL, Kuca K, Stourac P, Valis M (2016) Proteomic analysis of cerebrospinal fluid for relapsing-remitting multiple sclerosis and clinically isolated syndrome. Biomed Rep 5(1):35–40
3. Holtta M, Minthon L, Hansson O, Holmen-Larsson J, Pike I, Ward M, Kuhn K, Ruetschi U, Zetterberg H, Blennow K, Gobom J (2015) An integrated workflow for multiplex CSF proteomics and peptidomics-identification of candidate cerebrospinal fluid biomarkers of Alzheimer's disease. J Proteome Res 14 (2):654–663
4. Batth TS, Francavilla C, Olsen JV (2014) Off-line high-pH reversed-phase fractionation for in-depth phosphoproteomics. J Proteome Res 13(12):6176–6186
5. Yang F, Shen Y, Camp DG 2nd, Smith RD (2012) High-pH reversed-phase chromatography with fraction concatenation for 2D proteomic analysis. Expert Rev Proteomics 9 (2):129–134

Chapter 12

Brain Biomarkers: Follow-Up of RNA Expression Discovery Approach: CSF Assays for Neurogranin, SNAP-25, and VILIP-1

Elizabeth M. Herries, Nancy Brada, Courtney L. Sutphen, Anne M. Fagan, and Jack H. Ladenson

Abstract

Here, we describe our methods for identifying biomarkers of neurological disease that have the potential to evaluate patients with probable Alzheimer's disease (AD) and efficacy of novel therapeutics. We report a follow-up to our RNA expression discovery approach in mice to identify potential brain biomarkers. The gene expression approach identified 26 genes as having reasonable abundance and specificity and with human homologs known at the time. This chapter describes our follow-up in developing and evaluating the assays for the proteins expressed by these genes. We then present validated working assays for three analytes: VILIP-1 (visinin-like protein 1), a robust marker of neurodegeneration; synaptosomal-associated protein-25 (SNAP-25) and two different assays for neurogranin (Ng), as biomarkers of synaptic pathology, that we have utilized for studying AD, and which have demonstrated diagnostic utility.

Key words Cerebrospinal fluid, Biomarker, Alzheimer's disease, Neurogranin, SNAP-25, VILIP-1, Single Molecule Counting™ immunoassays

Abbreviations

AD	Alzheimer's disease
CSF	Cerebrospinal fluid
CV	Coefficient of variation
GST	Glutathione S-transferase
M.I.	Myocardial infarction
mAb	Monoclonal antibody
MPs	Magnetic microparticles
SMC™	Single Molecule Counting

Charlotte E. Teunissen and Henrik Zetterberg (eds.), *Cerebrospinal Fluid Biomarkers*, Neuromethods, vol. 168,
https://doi.org/10.1007/978-1-0716-1319-1_12,

1 Introduction

Stroke, Alzheimer's disease (AD), and other neurological diseases are major health problems. For some of the neurologic disorders, there are empiric therapies but for others, such as AD, there is currently no effective treatment. When we started the work described in this chapter, we had the original goal of developing assays in blood that would relate to brain damage akin to our previous work with heart damage and Myocardial Infarction (MI) [1]. Stroke was the original target but most of the clinical effort has related to AD, as will be described. When we started our work with brain biomarkers around 2002, it was known that amyloid plaques and Tau deposition in neurofibrillary tangles were the pathological hallmarks of AD, and the measurement of β-amyloid (Aβ) and Tau and phosphorylated Tau (pTau) in CSF were the only established biomarkers. We were aware that these were targets of potential therapy and so additional biomarkers could be valuable. We sought a discovery approach that would provide identification, abundance, and specificity of potential biomarkers. The two promising approaches to this problem were proteomics or RNA expression.

We chose RNA expression as there was a facile screening approach available for screening RNA in mouse organs (Affymetrix, Santa Clara, CA) with which we had experience, and we had no experience with proteomics and mass spectrometry.

We tested this approach by measuring gene expression in mouse heart and other organs. This clearly identified troponin as heart specific and previously used tests such as CK-MB and LD isoenzymes as not heart specific [2]. Had this knowledge been available 20 years ago, it would have saved many years of sometimes erroneous evaluation using only semi-specific tests for MI [1].

We then compared the RNA expression in brain and a number of other organs in C57Bl/6 mice and selected highly expressed genes that had at least a ten-fold higher expression in brain than other major organs. We checked for human homologs of the mouse proteins and further selected for proteins of relative low molecular weight (MW) (<70,000 daltons), rationalizing that they may be more prone to cross the blood–brain barrier (BBB) [3].

Twenty-nine genes were identified as having reasonable abundance and specificity and 26 had human homologs known at the time. Five proteins which had been studied as brain biomarkers; myelin basic protein (MBP), neuron-specific enolase (NSE), protein S-100B, glial fibrillary acidic protein (GFAP), and Tau did not meet our criteria for high basal gene expression in the brain. Twenty of the remaining 24 proteins were <70,000 daltons (Table 12.1).

Table 12.1
Proteins coded by genes with specific brain gene expression (Adapted from Ref. 3)

Protein name (Human)	Molecular weight (kDa)	Microarray units (RNA expression mouse)	Gene name	Antibodies and assays developed and epitope shown if known
GABA (A) receptor subunit γ2	54.2 (467 AAs)	10,620	GABRG2	Transmembrane 5 subunit protein. Did not try to make antibodies
Glutamate decarboxylase 1 brain M67,000 (GAD67)	66.9 (594 AAs)	20,931	GAD1	Capture 7A12 (4635) mAb hamster (A10-T22) Detection epitope 9A10 (WU-Ag-5-1.1) (T91-A100)
Glycine receptor, subunit beta	56.1 (497 AAs)	14,726	GLRB	Pentamer of alpha, beta subunits
Alpha-internexin	55.4 (499 AAs)	12,582	INA	Still being worked on; current available commercial assay has stated sensitivities of ~0.2 ng/mL
MOBP (Myelin-associated oligodendrocyte basic protein)	21.0 (183 AAs)	20,079	MOBP	Validation of assay in progress
Myelin proteolipid protein (lipophilin)	30.1 (277 AAs)	13,009	PLP1	Eliminated based on six cysteine and four transmembrane domains, inability to dissociate in SDS, and a smear on a Western blot. It is a multiple crossing transmembrane lipoprotein. Did not attempt to make antibodies
Neurensin-1	21.5 (195 AAs)	14,185	VMP	Did not try to make antibodies
Neurogranin (Ng)	7.6 (78 AAs)	10,322	NRGN	Assay developed with affinity purified rabbit antibodies Capture Rabbit 10055-P4793 (G49-A67) Detection Rabbit 10055—P4794 (A8–A30) Monoclonal antibody assay also developed Capture mAb (WU-Ag-7.1.1) 14D03 (D15-N27) Detection mAb (WU-Ag-11-5W) 10E11 (A28-A42)
Neuronatin	9.2 (81 AAs)	12,078	NNAT	Abundant in 18–24 week fetal brain. Minimal amounts in adult

(continued)

Table 12.1
(continued)

Protein name (Human)	Molecular weight (kDa)	Microarray units (RNA expression mouse)	Gene name	Antibodies and assays developed and epitope shown if known
Neuroserpin	46.4 (410 AAs)	12,309	SERPINI1	Assay developed. Capture (WU-Ag-8A-1.1) 8H3 (not mapped) Detection mAb 9G1 (N233-L242)
Noelin	55.3 (485 AAs)	13,884	OLFM1	Not yet worked up
Synaptobrevin-2	12.7 (116 AAs)	10,586	VAMP2	Synaptosome component
SNAP-25 (synaptosomal-associated protein)	23.3 (206 AAs)	25,438	SNAP-25	Assay uses (WU-Ag-4-1.1) 6H07 for capture (E151-A164) and (WU-Ag-4-1.1) 9E11 for detection (E13-A22)
Synaptolagmin-1	47.6 (422 AAs)	15,650	SYT1	Transmembrane homotetramer lipoprotein. Did not try to raise antibodies
Syndapin-1 (Protein kinase C and casein substrate in neurons1)	51.0 (444 AAs)	13,143	PACSIN1	Protein similar in rat and mouse but somewhat different in human. For assay used capture mAb 23G4 (T220-E232) and detection 1A7 (Q76-G85). All react with Pacsin-1 not Pacsin 2 or 3
Tubulin, beta-4A chain	49.6 (444 AAs)	15,537	Tubb4A	Failed to obtain antibodies
Vesicular inhibitory amino acid transporter	57.4 (525 AAs)	12,534	SLC32A1	Multi-crossing transmembrane protein. Did not try to make antibodies
Visinin-like Protein-1 (VILIP-1)	22.1 (191 AAs)	25,838	VSNL1	Assay uses capture 3A8.1 (4399) (S96-Y108) Detection mAb 2B9.3 (4403) (F55-D73) Prior assays (see Table 12.2) used affinity purified rabbit 4729 or sheep 22 test bleed affinity purified
Zinc finger protein (ZIC-1)	48.3 (447 AAs)	13,143	ZIC1	No effective antibodies developed

(continued)

Table 12.1 (continued)

Protein name (Human)	Molecular weight (kDa)	Microarray units (RNA expression mouse)	Gene name	Antibodies and assays developed and epitope shown if known
Zygin-1 (Fasciculation and elongation protein) Zeta 1	45.1 (392 AAs)	19,864	FEZ1	Assay developed but not effective for neurological damage Capture mAb (4554) 1G4 (P23-Q28) Detection mAb (4563) 4G3.1 (S7-F12) LLoQ-14 pg/mL

We confirmed strong protein expression of the gene products by Western blot (WB) with either commercial antibodies or antibodies that we developed. Thirteen of the proteins were analyzed further by WB analysis of a normal human tissue array (Geno Technologies, Inc., St. Louis, MO). Those which showed reasonable expression and specificity for brain compared to other organs were selected for further evaluation.

We expressed the proteins or immunogenic peptides contained in them, generally coupled to glutathione S-transferase (GST), and immunized Swiss-Webster mice or Armenian hamsters for preparation of monoclonal antibodies, and also rabbit or sheep for the more rapid development of polyclonal antibodies. The epitopes of promising antibodies were mapped by immunostaining, peptide reactivity, and/or ABIMED spot peptide arrays prepared at the MIT Biopolymers Lab (Cambridge MA). Each spot comprised a 10-mer peptide anchored to the membrane at the peptide's C-Terminus generally with a 1 or 2-residue offset. A summary of our work with the proteins identified as candidates in reference [3] is shown in Table 12.1.

The general scheme we followed was to utilize antigen-coated microtiter plates to select antibodies. Typically, polyclonal antibodies were available well before monoclonal ones which allowed us to assess the possible utility of the analyte more rapidly. Rather than separately remove anti-GST antibodies and then make affinity purified ones, we developed a one-step affinity purification system that was rapid and effective [17]. In some cases, we continued to use the affinity purified polyclonal antibodies after we had monoclonal ones if the assay using them was more sensitive as we could not predict the concentrations in CSF or blood for the disorders we were assessing. Protein expression could be different than gene expression predictions, and post-translational changes could affect the epitope recognized by the antibodies.

We originally developed the assays using electrochemiluminescence detection on a Meso Scale Discovery (MSD) platform. Later, we utilized Single Molecule Counting™ (SMC™) detection [18] with the Erenna immunoassay instrument from Singulex (Alameda, CA) (now available from Millipore Sigma, Burlington, MA) which provided two- to six-fold greater sensitivity. After performing pilot clinical studies, we determined that greater sensitivity was not needed for some of the analytes, but generally we continued to use the more sensitive SMC™ Erenna platform which reduced sample volume requirements. We also found in a later pilot study that the Single Molecule Array (SIMOA) assay using the Simoa HD-1 instrument (Quanterix, Lexington, MA) [19] could give similar sensitivities for VILIP-1.

We first evaluated two-site immunoassays in human postmortem CSF and CSF from a rat model of stroke for the assays which shared the antibody epitopes between rat and human. When a promising pair of antibodies for capture and detection were identified, we evaluated assay sensitivity, precision, dilutional linearity, and spike recovery using different buffers and assay conditions. We put particular emphasis on dilutional linearity as we found this was essential to get reproducible dose response with the analytes. We initially concentrated our work on the protein coded for by the highest RNA expression, VILIP-1. This chapter will review our experience with the candidate proteins in their alphabetical order rather than the degree of gene expression used in reference [3], and then present the validated working assays we have utilized for studying AD; Ng, SNAP-25, and VILIP-1.

2 Glutamate Dehydrogenase, MW 67 (GAD 67, GAD 1)

Antibodies were raised to His-tagged GAD 67 in Swiss-Webster mice. An assay was constructed with mAb 7A12 as a capture antibody (epitope A10-T22) and mAb 9A10 (T91-A100) as a detection antibody with little reactivity with GAD25. We performed analysis of GAD 67 on 59 CSF samples kindly sent to us by Dr. Kaj Blennow at the University of Gothenburg, Sweden. GAD 67 was undetectable in 26 controls and only at low concentrations in 5 of the AD samples. We have not yet pursued this assay further.

3 Myelin-Associated Oligodendrocyte Basic Protein (MOBP)

MOBP is found in myelin and is one of the two white matter proteins identified in our RNA expression screen (Table 12.1), the other being the gene *PLP-1* (protein, lipophilin). White matter involvement in AD has been proposed [20]. We first developed antibodies with GST-MOBP that we made or purchased from

Abnova (Taipei, Taiwan), and lipophilin was not promising for developing an assay as noted in Table 12.1. The cleavage of GST-MOBP proved to be very difficult and is still not yet fully resolved. We tried cloning into pGex-4T1 or pGex-6P1 but had difficulty with both due to formation of precipitates. After trying numerous strategies, including addition of the detergent, beta-octyl-glucoside, we still had only minor success.

As we mapped the epitopes of the antibodies we obtained with GST-MOBP, we made peptides corresponding to epitopes with which they reacted and used them to affinity purify the polyclonal antibodies or immune deplete them. We plan to make antibodies to the C-terminal part of the molecule with characteristics similar to sc-14520 which we epitope mapped but is no longer available from Santa Cruz Biotechnology, Inc. (Dallas, TX).

After trying various antibody pairs for both capture and detection with mixed results, we now believe that there is a difference (possibly conformational) between GST-MOBP and MOBP found in human brain lysate and are developing a secondary standard of CSF or brain lysate for assay evaluation. We did not test our CSF from the rat stroke model due to the low amount of white matter in rat brain.

4 Neurogranin (Ng)

Another one of the brain-specific proteins identified by RNA expression which was of high interest was the one with the lowest MW, neurogranin (Ng), 7.6 kDa (protein kinase C substrate, Rc3).

We immunized mice and rabbits with two different Ng recombinant proteins differing in the linker area between GST and Ng; 4T-1 (thrombin site for cleavage) and 6P1 (cleavage site for the PreScission enzyme). These were quite suitable for getting antibody response, but we had a great deal of difficulty getting sufficient Ng after cleavage of the GST. We therefore had Ng synthesized by AAPPTec (Louisville, KY) and quantitated by amino acid analysis (AAA). We prefer AAA for setting standard concentrations in general, as A280 will not work if there are no tryptophan or tyrosine residues, which is the case with Ng. We obtained good antibody responses in rabbit (R10055). We had previously identified peptides P-4793 (amino acids G49-G60) and P-4794 (amino acids S10-D23) which might be good immunogens. We affinity purified R10055 separately with P-4793 and P-4794 and developed a "two-site" immunoassay; capture antibody (G49-67, not G49-C60 as we previously reported [21]), and antibody (A8-A30, not S11-D23 as previously reported [21]), for detection (this assay also works in CSF of rats and mice) and later some effective monoclonal antibodies.

Neuromodulin, which has similarities in structure to Ng and is elevated in AD [22], was not reactive by WB with either of the affinity purified rabbit antibodies or with mAb 14D3 (D15-N27) but was with another of our monoclonal antibodies, 10E11 (A28-A42).

Using either the Mesoscale Discovery or the Erenna, very sensitive immunoassays using affinity purified rabbit antibodies were developed, and very high concentrations were obtained in CSF from our rat stroke model. Furthermore, in our hands, a variety of clinical studies in CSF have shown significant increases in Ng in individuals with AD compared to controls [8, 15, 16, 21, 23]. While these publications employ the assay in which urea was used for elution of the Alexa® Fluor coupled detection antibody, we now use glycine which correlates highly with urea elution, but gives concentrations ~25% higher. This assay using glycine was first reported in a study of brain biomarkers in obstructive sleep apnea where decreases in Ng were seen compared to a control group [13].

Here, we also describe a new 2-monoclonal antibody Ng assay using 14D3 for capture and 10E11 for detection. While mAb 10E11 reacted with neuromodulin on WB, we had no cross-reactivity in our assay with up to 50 ng/mL neuromodulin added, indicating that there was no nonspecific sticking of neuromodulin to the magnetic microparticles, as the capture antibody does not bind neuromodulin. Comparing the two-mAb assay to the rabbit assay (urea elution), using CSF samples from some of our prior clinical studies, we observed that the two assays correlated well, while concentrations for the monoclonal assay were about 30% lower ($R^2 = 0.88$).

The assays for Ng detection using rabbit (P4793/P4794) affinity purified antibodies and monoclonal (14D3/10E11) antibodies, both using glycine elution, will be described later in this chapter.

Other groups have also published on Ng. Davidsson and Blennow [24] identified a significant decrease in synaptic proteins in brain including Ng in AD but not in vascular dementia at postmortem examination consistent with known synaptic involvement in AD. Kvartsberg et al. [25, 26] reported elevations of Ng in CSF using a capture mAb Ng7, epitope R53-A64 [27] with a rabbit antibody (from Upstate Biotechnology, Lake Placid, NY (per personal communication with Dr. Kaj Blennow), now sold as 07-425 by Millipore Sigma) for detection. Interestingly, when brain lysate was extracted with these antibodies separately, mass spectrometry showed intact Ng in brain tissue but not in CSF [25]. Higher Ng concentrations were reported in plasma when compared with CSF, with clear separation of AD and controls observed only in CSF [26, 27].

Another assay for Ng was described by DeVos et al. [28] which used mAb ADxNg C12 for capture (epitope R51-A66) and mAb

ADxN6 CTI for detection (epitope G62-P75). The mAb ADxNgCI2 was indicated to be directed to the same epitope as Ng7 but had higher affinity. This antibody was chosen on the basis that its epitope seemed more abundant in CSF than the intact C-terminus. The standard for the assay was Ng truncated at P75.

There are a number of clinical studies utilizing all the assays noted above. A head-to-head comparison of three of these assays (referred to as WashU, UGOT, and ADX) was recently published [23]. While there were differences in absolute values obtained by each of the assays, all three discriminated controls and patients with AD, and measured values correlated well. The assays tested in this comparison study utilized the following antibodies:

(a) UGOT Ng7 capture (R53-A64 [27], rabbit antibody 07-425 Millipore Sigma mapped to G65-D78 for detection) [27]. This group has since reported a new double monoclonal assay that uses Ng36 (immunogen was KLH-conjugated peptide G63-P75) for capture and Ng2 (epitope R52-G63) for detection [29].

(b) ADX mAb ADxNGCL2 for capture (R51-A66) and mAb ADNGCTI for detection (G62-P75).

(c) WashU affinity purified rabbit antibodies, P4793, capture (G49-A67) and P-4794 for detection (A8-A30). Note: Reference [23] reports P4793 epitope as (G48-C60) and P4794 epitope as (S10-D23); the correct epitopes are given above.

5 Neuroserpin (Serpinl1)

Using Abnova GST-neuroserpin as immunogen; monoclonal antibodies were developed, mAb 8H3 (not mapped) for capture and 9G04 (N233-L242) for detection. Other antibodies were also made, and some reacted with human but not mouse or rat neuroserpin by WB. Full validation studies are not yet performed.

6 Synaptosomal-Associated Protein, MW 25,000 (SNAP-25)

SNAP-25 was cloned into pGexT-3 and used for immunization. Two of the monoclonal antibodies were used for our assay; capture antibody 6H07, epitope E151-A164 and detection antibody 9E11, epitope E13-A22. The lower limit of quantification [LLoQ] (defined in Subheading 10.5.2, **step 3**) for the assay is 0.08 pg/mL. The gene was the second most highly expressed in normal mouse brain and only weakly in any other organ. Useful antibodies were also obtained from sheep and rabbits. SNAP-25 is highly conserved in human, rat, mouse, and chicken, so the assay can be readily utilized in other species. The SNAP-25 standard was obtained from Cell Sciences (Newburyport, MA). We have published three papers regarding SNAP-25 in AD and obstructive sleep apnea [13, 15, 16]. Others have also reported an assay for SNAP-25 [30].

7 (Syndapin-1) Protein Kinase C and Casein Kinase Substrate in Neurons 1

Antibodies were raised by immunization with GST-syndapin-1 from Abnova. Our assay utilized a standard obtained from Abnova and two monoclonal antibodies, 23G4 for capture (T220-E232) and 1A7 (Q76-G85) for detection. The antibodies do not cross-react with syndapin 2 or 3. The assay has an LLoQ of 3.5 pg/mL. Initially, we found considerably less syndapin-1 than VILIP-1 in CSF drained from patients with hemolytic stroke (MSD platform). As the assay was further refined, this was no longer true, but detailed follow-up has not yet been performed. It is of interest that syndapin-1 has been reported to interact with Tau protein [31].

8 Visinin-Like Protein-1 (VILIP-1)

VILIP-1 is a highly conserved protein, identical in human, mouse, rat, bovine, and chicken. It had the highest specific gene expression in mice and so we worked with this protein first. Antibodies were raised to GST-VILIP-1 expressed in E. Coli using PGex vectors.

VILIP-1 was quite high in the CSF of a rat model of stroke 24 h post-surgery but had little change before that. In a traumatic brain injury (TBI) model in mice, the highest concentrations of VILIP-1 were found when there was evidence of BBB breakdown.

In an unpublished study of human participants with acute neurologic deficits suggestive of a stroke, the final discharge diagnosis was ischemic stroke ($n = 64$), TIA, or stroke mimics ($n = 56$). Significant increases in plasma VILIP-1 were most pronounced at 72 h (Fig. 12.1, unpublished), which is consistent with the known breakdown in the BBB following ischemic stroke, and makes it unlikely that VILIP-1 could be a consistent early marker of stroke in blood. We found somewhat similar results in a small pilot study of patients with head injury in the emergency department. In 25 patients with head injury, clear elevation of serum VILIP-1 levels was only found in the two subjects that had evidence of brain hemorrhage by CT scan. These results are consistent with those found for plasma Glial Fibrillary Acidic Protein (GFAP) [32] and suggest that some breakdown in the BBB such as hemorrhage may be necessary for most brain proteins to accumulate in the blood.

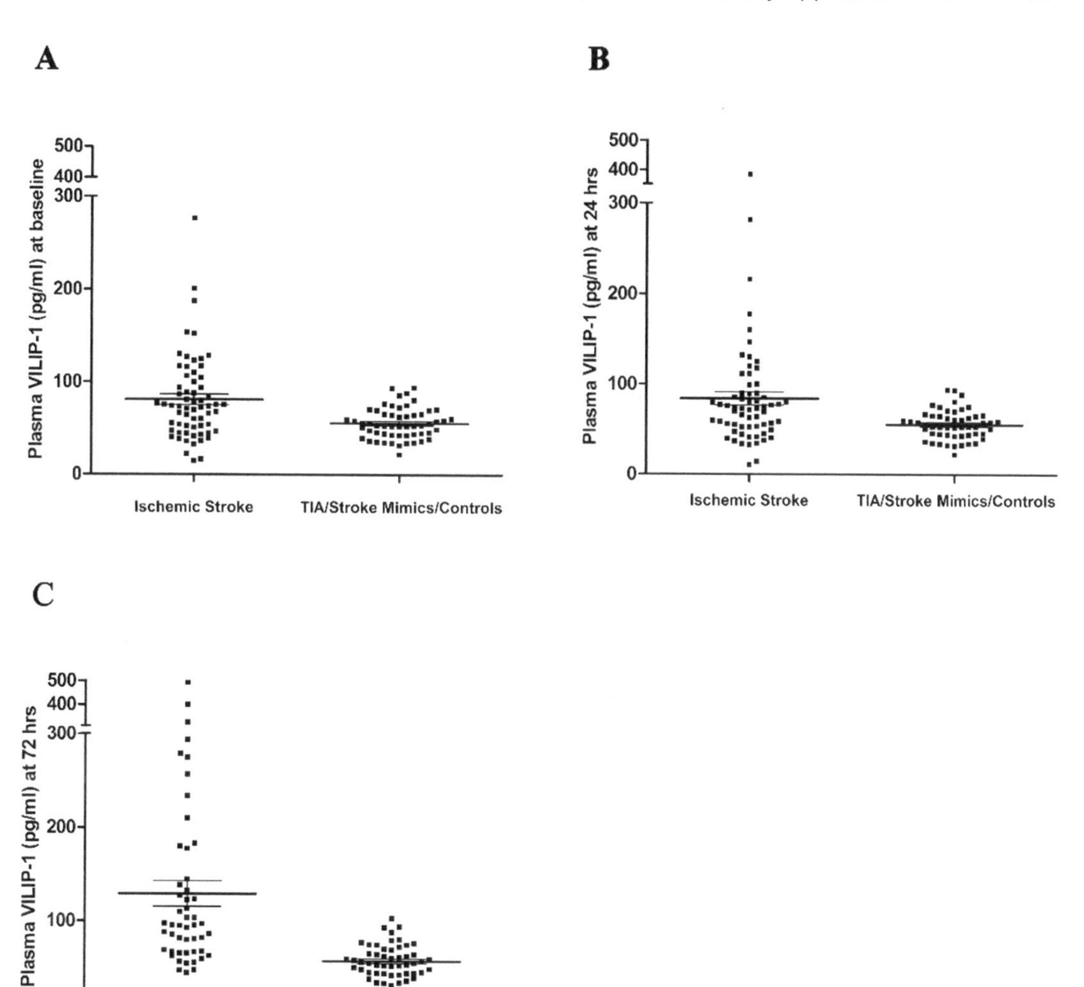

Fig. 12.1 Plasma Biomarker Concentrations in Ischemic Stroke Compared to TIAs, Stroke Mimics and Controls at Baseline, 24 h and 72 h. (**a**) Participants with ischemic stroke had higher plasma VILIP-1 concentrations (mean $\pm$ SE; 81.1 pg/mL $\pm$ 5.7, $N = 64$) at baseline compared to the combined cohorts of TIAs, stroke mimics, and controls (55.9 pg/mL $\pm$ 2.2, $N = 56$) ($P = 0.0002$). (**b**) Mean ($\pm$SE) plasma VILIP-1 concentrations at 24 h were higher in ischemic stroke (84.0 pg/mL $\pm$ 7.4; $N = 64$) compared to the combined cohorts of TIAs, stroke mimics, and controls (55.6 pg/mL $\pm$ 2.2; $N = 53$) ($P = 0.001$). (**c**) Similarly, mean ($\pm$SE) plasma VILIP-1 concentrations at 72 h were higher in ischemic stroke (129.1 pg/mL $\pm$ 13.6; $N = 51$) compared to the combined cohorts of TIAs, stroke mimics, and controls (57.6 pg/mL $\pm$ 2.5; $N = 53$] ($P < 0.0001$)

In addition, we assessed VILIP-1 in the CSF of individuals with AD. In our initial study of 33 individuals with AD and 24 healthy controls, VILIP-1 was significantly elevated in disease [4]. We also assessed VILIP-1 in plasma in a small study of 64 individuals with AD compared with 149 cognitively normal controls (Fig. 12.2, adapted from Supplemental data, [5]). While there was a statistically significant difference between both very mildly impaired (Clinical Dementia Rating [CDR] 0.5) and mildly impaired (CDR 1) individuals and cognitively normal controls, there was

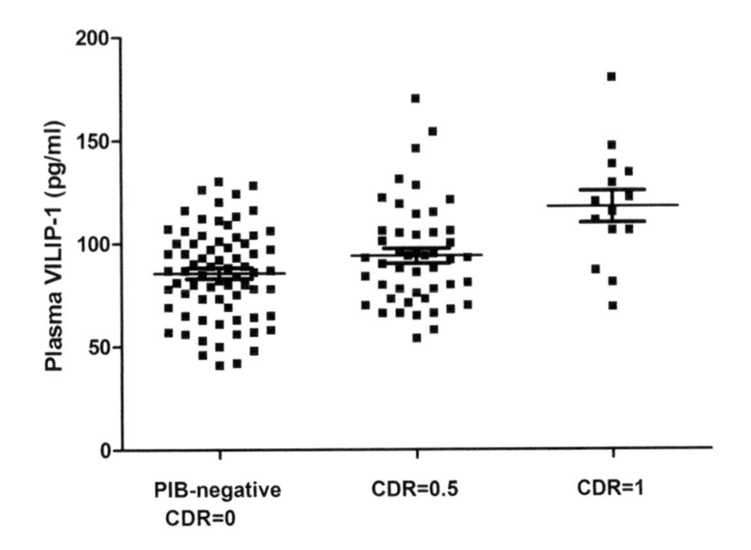

Fig. 12.2 Plasma VILIP-1 concentrations by clinical status category in Cognitively Normal Controls (CNC) and AD. Mean (±SE) plasma VILIP-1 concentrations were significantly elevated in individuals with mild dementia (Clinical Dementia Rating [CDR 1]) (118 ± 7.6 pg/mL, $n = 14$) compared to amyloid-negative (PIB-negative) CNC (86 ± 2.6 pg/mL, $n = 72$) ($p < 0.0001$) and very mildly demented individuals (CDR 0.5) (94 ± 3.6 pg/mL, $n = 48$) ($p = 0.0038$). Mean plasma VILIP-1 concentrations were higher in CDR 0.5 compared to PIB-negative CNC ($p = 0.054$). The moderately demented (CDR 2) group included only two individuals (not shown). One-way ANOVA with Bonferroni's correction was performed for all group comparisons

not a clear separation between these groups so we preferentially pursued VILIP-1 in CSF [4–7, 9–11, 14].

Initially, we developed the VILIP-1 assay using a monoclonal capture antibody, 3A8.1, (epitope S96-Y108) and affinity purified rabbit 4729 as detection antibody. mAb 3A8.1 did not react with other EF hand domain neuronal calcium sensor [NCS] human proteins; VILIP-3, neurocalcin, or recoverin. Polyclonal sheep antibodies were also tested as a means to increase antibody supply. Both test and final sheep bleeds were mapped and each had 6 non-overlapping epitopes, albeit not the same in the test and final bleeds. We replaced affinity purified rabbit 4729 with affinity purified sheep 22 test bleed. When used as the detection antibody, sheep 22 test bleed is calcium dependent, so calcium was added to all assay buffers.

We originally standardized the assay with VILIP-1 from GST-VILIP-1, with the GST cleaved by thrombin ("cut" VILIP-1) and the concentration assigned by A280. We realized that we had some variation using A280 for setting the standard concentration and changed to using amino acid analysis (AAA). In three different standard preparations, the AAA concentrations were 46% lower than those calculated by A280. It may be important to use laboratories for amino acid analysis that assess bovine serum albumin (BSA) from the National Institute of Standards in the USA (BSA-NIST) routinely or run it on request as it has been noted that there can be variation in concentrations set by amino acid analysis sent to different laboratories (Kaj Blennow, personal communication).

As an alternative for the sheep 22 polyclonal detection antibody, we looked again at one of our previously developed monoclonal antibodies, 2B9.3 (epitope F55-D73) which reacted with VILIP-1, VILIP-3, and neurocalcin but not recoverin. This had originally given us poor performance with recombinant cut VILIP-1. We obtained VILIP-1 generated with a different expression vector from Proteos (Kalamazoo, MI) that came as both monomer and dimer conformations. We utilized the monomer as our standard after we determined that VILIP-1 was monomeric in our human CSF control samples. We also noted that our recombinant cut VILIP-1 was mostly dimer. VILIP-1 values in CSF with the two assays correlated well and gave similar concentrations with calcium in the assay buffer. This two-monoclonal antibody assay described later is now utilized in all our studies [14]. It has no reactivity with VILIP-3, neurocalcin or recoverin. Table 12.2 describes the differences in regard to the method of standardizing, the detection antibody used (mAb 3A8 was used as the capture antibody in all our clinical studies with VILIP) and whether urea or glycine was used for elution of the labeled detection antibody, with references to specific publications where each assay iteration was used [3–16]. To elute the detection antibody for quantification on the Erenna instrument, we initially utilized urea but later changed to glycine which gave standard curves with greater slopes and more stable concentrations, with elution complete within 5 min. We now use glycine for elution in all our assays.

We also performed some pilot studies in samples obtained from patients with other disorders:

(a) Traumatic brain injury (TBI)—As noted above, clear elevations were only found in ER patients or animals that had evidence of hemorrhage which would disrupt the BBB.

(b) Amyotrophic Lateral Sclerosis (ALS)—In the three patients tested, VILIP-1 was in the range of the clinically normal subjects from our AD studies.

(c) Little to no changes in VILIP-1 were observed in Multiple Sclerosis.

(d) Parkinson's Disease (PD)—There was no correlation of VILIP-1 with time of diagnosis, but elevations were found in a few patients.

(e) Hypoxia-ischemia or stroke in neonates. VILIP-1 is quite high in CSF of neonates, and no effect of hypoxia was apparent.

Table 12.2
Publications related to VILIP-1 with assay variations noted

Reference	Standardization (A280 or AAA)	Detection antibody (mAb 3A8 for capture) All Erenna platform unless noted	Elution agent in buffer
Laterza, et al. (2006) Clin Chem 52, 1713–1721 [3]	A280	Rabbit 3471 (MSD platform)	Not applicable
Lee, et al. (2008) Clin Chem 54, 1617–1623 [4]	A280	Rabbit 3471 (MSD platform)	Not applicable
Tarawneh, et al. (2011) Ann Neurol 70, 274–285 (~300 participants) [5]	A280	Sheep 22 *all the assays with sheep 22 were from test bleed	Urea for CSF Glycine for plasma
Tarawneh, et al. (2012) Neurology 78, 709–719 (60 participants; subset of 300) [6]	A280	Sheep 22	Urea
Tarawneh, et al. (2015) JAMA Neurol 72, 656–665 (87 participants) [7]	A280	Sheep 22	Urea
Tarawneh, et al. (2016) JAMA Neurol 73, 561–571 (302 participants) [8]	A280	Sheep 22	Glycine
Fagan, et al. (2014) Sci Translat Med 6, 223ra230 [9]	AAA	Sheep 22	Glycine
Kester, et al. (2015) Alzheimer's Res Therapy 7, 59 [10]	AAA	Sheep 22	Glycine
Sutphen, et al. (2015) JAMA Neurol 72, 1029–42 [11]	AAA	Sheep 22	Glycine
Shahim, et al. (2015) Brain Injury 29, 872–6 [12]	AAA	Sheep 22	Glycine
Ju, et al. (2016) Ann Neurol 80, 154–9 [13]	AAA	mAb 2B9	Glycine
Crimmins, et al. (2017) Clin Chem 63, 603–604 [14]	AAA	mAb 2B9	Glycine
Sutphen et al. (2018) Alzheimers Dement. 14, 869–879 [15]	AAA	Sheep 22	Glycine
Schindler et al. (2019) Alzheimers Dement. 15, 655–665 [16]	AAA	mAb 2B9	Glycine

(f) Studies of samples obtained from patients with autoimmune-mediated encephalitis (AIME) are ongoing with Dr. Gregory Day and his colleagues.

To the best of our knowledge, only one other assay for VILIP-1 has been published [33], and it is available from BioVendor (Brno, Czech Republic).

9 Zygin

Assays for zygin with various antibody combinations worked well with standards, but the only format which detected zygin in post-mortem CSF utilized two monoclonal antibodies with epitopes close together; 1G4.4 (P23-Q28) for capture and 4G3.1 (S7-F12) for detection and an LLoQ of 1 pg/mL. Quite high concentrations were obtained in postmortem CSF (1300 pg/mL), but not in our other CSF samples. After further assessment, we decided that degradation was a probable issue and stopped working with the assay.

While the antibodies described here are currently not yet commercially available, discussion concerning their use can be made with Washington University Office of Technology Management or with the corresponding authors.

10 CSF Immunoassay Procedures for Neurogranin, SNAP-25, and VILIP-1

10.1 Assay Methods Overview

Our immunoassays for Ng, VILIP-1 and SNAP-25 employ a quantitative fluorescent sandwich technique with detection via a Single Molecule Counting (SMC™) method originally marketed by Singulex as their Erenna® system [18] (now available from Millipore Sigma), to measure analytes in human CSF using the antibodies and assay conditions that were developed in the Ladenson Laboratory at Washington University.

The general scheme is a two-site immunoassay with a capture antibody bound to magnetic microparticles and a detection antibody labeled with Alexa® Fluor 647. An Erenna instrument is used to measure the amount of Alexa® Fluor 647 labeled "detection" antibody eluted from analyte bound to microparticles. For the capture step, "capture" antibody-coated microparticles, standards, and diluted CSF samples are combined into microplate wells for two hours. Unbound molecules are washed away. Alexa® Fluor-labeled detection antibody is then added and binds to the analyte captured onto the microparticles. Following another wash step, the microparticles are transferred to a clean plate, and the labeled detection antibodies are eluted from the microparticles and then read on the Erenna.

The advantages of the technique are high sensitivity, small sample volumes and a large linear range. The disadvantage is the need for a specialized instrument. We believe that the assays could be adapted to other sensitive immunoassay systems available for research and clinical laboratories. While assays for CSF are described here, the assays can also be adapted for analysis of blood samples.

The following sections describe basic requirements for analyzing the three analytes, including materials: equipment, supplies, required reagents, chemicals, preparation of buffers, capture and detection antibody reagents, quality control (QC) samples, and standards. These are followed by assay preparation instructions and the assay procedure for all of the assays.

Note: As a precaution, we routinely replace reagents after 6 months. However, reagents can potentially be used longer if assay characteristics, e.g., slope, QC results, are unchanged.

10.1.1 Assay Flowchart (Fig. 12.3)

10.2 Materials

Specialized laboratory equipment required for the assays includes:

10.2.1 Equipment

1. Singulex Erenna® (or Millipore Sigma [St Louis MO] SMCxPRO™).
2. Microplate "orbital" shaker [VWR PN 12620-926] capable of slow acceleration to the speed setting.

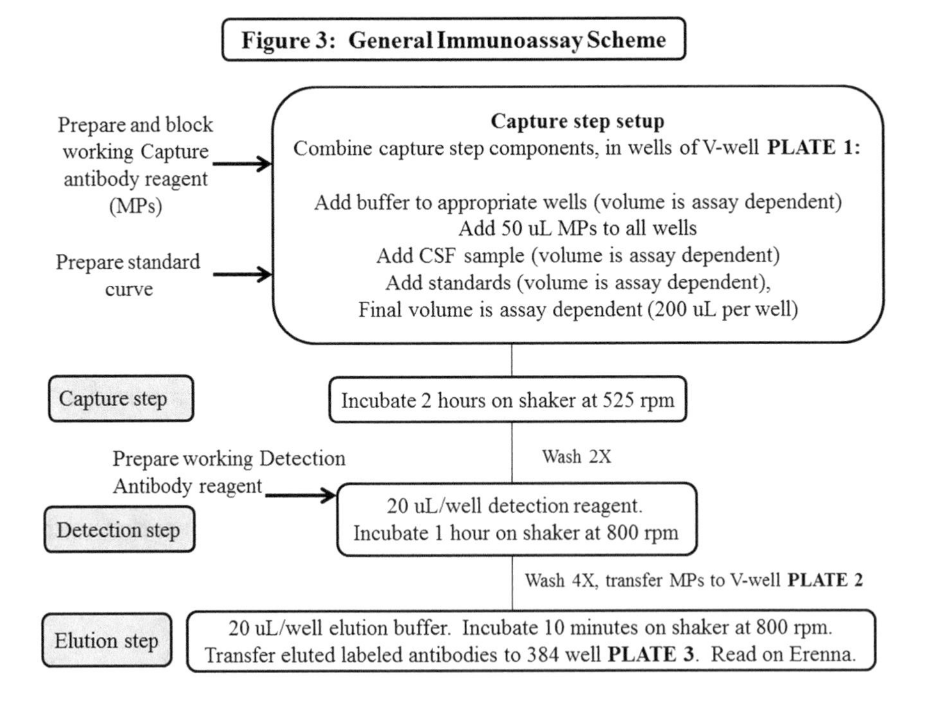
Figure 3: General Immunoassay Scheme
Prepare and block working Capture antibody reagent (MPs)
Prepare standard curve
Capture step setup
Combine capture step components, in wells of V-well PLATE 1:
Add buffer to appropriate wells (volume is assay dependent)
Add 50 uL MPs to all wells
Add CSF sample (volume is assay dependent)
Add standards (volume is assay dependent),
Final volume is assay dependent (200 uL per well)
Capture step
Incubate 2 hours on shaker at 525 rpm
Wash 2X
Prepare working Detection Antibody reagent
Detection step
20 uL/well detection reagent.
Incubate 1 hour on shaker at 800 rpm
Wash 4X, transfer MPs to V-well PLATE 2
Elution step
20 uL/well elution buffer. Incubate 10 minutes on shaker at 800 rpm.
Transfer eluted labeled antibodies to 384 well PLATE 3. Read on Erenna.

Fig. 12.3 Assay flowchart. General scheme for Ladenson Lab single molecule counting Neurogranin (monoclonal and rabbit polyclonal versions), SNAP-25, and VILIP-1 two-site immunoassays

3. Stat Fax® 2200 Microplate Incubator/Shaker [Midsci Inc., Valley Park MO, PN SF2200] (A second shaker is useful when running multiple assays in a day, as assay incubation steps use different mixing speeds). The Stat Fax unit has a light blocking cover and is well suited for detection and elution step mixing.
4. FluidX [Salt Lake City UT] X-Seal Manual Thermal Sealer (model HT121TS).
5. Centrifuge (with microplate carrier capable of speeds of 46 and 1900 ×*g*).
6. Refrigerated microcentrifuge (with rotor capable of speed of 11,000 ×*g*).
7. Vacuum pump (Welch model 2511B-01, Fisher Scientific part # 01-055-60).
8. Rotisserie Rotator (i.e., Labnet Mini-roller).

   Automated plate washer options include:

9. Tecan Hydroflex™ microplate washer (Millipore Sigma PN 95-0005-02).
10. Bio-Tek [Winooski VT] ELx405™ Microplate washer.

10.2.2 Supplies

Some non-standard laboratory supplies are required for the assays or their reagent preparation.

1. **Millipore Sigma**: 4titude Pierce Seal (heat sealing) Plate Foil (PN 02-01-0216-00).
2. **Thermo Fisher Inc [Rockford IL]**: DynaMag™ -96 side Skirted Magnet (PN 12027).
3. **Fisher Scientific [Pittsburgh PA]**:
   (a) Axygen 96-well V-well polypropylene plate (PN 14-222-240), **assay plates 1 and 2.**
   (b) MicroTube™ 8-tube strips (11-842-85).
   (c) Nunc 384-well polypropylene plate (PN 12-565-384), **Erenna analysis plate 3.**
   (d) Ultrafree MC–GV Durapore™ 0.2 μm centrifugal filters (PN UFC30GVNB).
   (e) Durapore 0.2 μm syringe filter (PN SLGV033RS).
   (f) Amicon Ultra-15 MWCO centrifugal filter units (PN UFC903024).
   (g) Nalgene Rapid-Flo 1 liter filter flask (PN 164-0020).

10.2.3 Reagents

1. **Erenna SMC™***reagentsfromMilliporeSigma*
   - Detection Antibody Reagent Labeling Kit (PN 03-0076-02).

- 10X System/Wash buffer with Proclin, 1 liter (PN 02-0111-03).
- Elution Buffer B (Glycine), 100 ml (PN 02-0297-00).
- Buffer C (1 M TRIS neutralization buffer for use with glycine elution buffer), 100 mL (PN 02-0298-00).
- Erenna® 20 Set Instrument Calibrators (PN 02-0046-02).

2. **Thermo Scientific**
   - EZ-Link Sulfo-NHS-LC-LC-Biotin (PN 21338).
   - Dynabeads MyOne Streptavidin T1(PN 65601), magnetic microparticles **for the Monoclonal Ng assay only.**
   - Dynabeads MyOne Streptavidin C1(PN 65001), magnetic microparticles **for SNAP-25, VILIP-1, and Rabbit Ng assays.**

10.2.4 Chemicals

Assay buffer components (equivalent chemicals can be used):

1. **Millipore Sigma**
   - Triton X-100 (T9284).
   - Trizma Base (T1504).
   - BSA (globulin-free, heat shock fraction) (A-7030).
   - 1 M Calcium Chloride solution (21115).
   - Hydrochloric Acid (H1758).
   - Borate-buffered saline tablets (PN 08059).
   - Sodium carbonate (S-7795).
   - Sodium bicarbonate (S-6014).
2. **Equitech-Bio [Kerrville TX]**:
   - Mouse IgG (SLM56).
   - Rabbit IgG (SLR56), both as lyophilized powder.
3. **Fisher Scientific**
   - Sodium chloride (BP358-212).
   - Sodium azide (BP922I).
4. **Thermo Scientific/Pierce Biotech**
   - Blocker Casein in TBS (37532).
   - BupH™ Phosphate-Buffered Saline [PBS] packs (PN 28372).

10.2.5 Buffers Used for All Assays

1. Triton X-100 (10%). Prepare ~10% solution (w/v) by weighing 1 g (of Triton X-100 into 9 mL deionized water. The ~10% solution is stable for 1 week.
2. Sodium azide (10%). Prepare 10% solution (w/v) by weighing 1 g sodium azide into 10 mL deionized water. The 10% solution is stable for 6 months. (CAUTION: this is a toxic carcinogen).

3. System buffer for Erenna instrument: 1 L of system buffer is prepared by diluting Erenna SMC™ 10X System/Wash Buffer (*see* Subheading 10.2.3, **step 1**) 1:10 in deionized water. The buffer is then filtered through a Nalgene Rapid Flo 0.2 micron filter flask, and thoroughly degassed by vacuum pump (*see* Subheading 10.2.1, **step 7**) for 15–30 min with stirring on a magnetic stir plate, prior to use. Attach to Erenna Analysis Buffer port. System Buffer is stable for 1 month.
4. Wash buffer for the assays: When using an automated plate washer, prepare 400 mL for a single assay plate or 700 mL for two assay plates (which includes additional volume for priming the washer's liquid lines) by diluting Erenna SMC™ 10X System/Wash buffer (*see* Subheading 10.2.3, **step 1**) 1:10 in deionized water. This is identical to the system buffer, except it is not filtered or degassed. It is utilized on the day it is prepared.

10.2.6 Buffers Used for Specific Assays

1. Two assays (monoclonal Ng and VILIP-1) use a single "Assay" buffer for standard curve preparation, sample dilution, preparation of capture antibody reagent, (i.e., Capture Antibody-Coated Magnetic Microparticles) and detection antibody reagent, (i.e., Labeled Detection Antibody). Two of the assays (Rabbit Ng and SNAP-25) employ two buffers: a "**Standard and Sample buffer**" for standard curve preparation and sample dilution and an "**Assay buffer**" for preparing Capture and Detection antibody reagents.
2. All **Assay buffers** contain mouse or rabbit IgG as appropriate to block nonspecific binding and as potential confounders for human antibodies to animal immunoglobulins in samples (e.g., human anti-mouse antibodies (HAMA) reactivity). Assay Capture antibody and Detection antibody reagents are prepared **only** in these IgG-containing assay buffers. Note: add IgG to buffer already containing Triton X-100, as it is more soluble with Triton X-100 present.
3. All buffers are sterile filtered using Durapore 0.2 μm syringe filters, at the time of preparation, with the exception of SNAP-25 assay buffers (which use a viscous casein buffer). Filtered buffers are stable for 1 month at 4 °C.
4. *Monoclonal Ng Assay Buffer.* 0.05 M TRIS, 0.15 M NaCl, 0.05% sodium azide, 0.25% Triton X-100, pH 8.1, to which is then added 2 mg/mL Mouse IgG. Filtered buffer (before addition of $CaCl_2$) is stable for 1 month at 4 °C. Add 0.08 mL of 1 M calcium chloride ($CaCl_2$) solution to 40 mL buffer (final concentration of $CaCl_2$ is 2 mM) at the time of the assay, as over time $CaCl_2$ will precipitate out of solution, causing it to

appear cloudy. All directions applying to this buffer in the future assume that $CaCl_2$ has been added. Forty mL buffer is required for one assay plate.

5. Rabbit Ng Assay Buffers
   - **Rabbit Ng Assay Base buffer**: 0.05 M TRIS, 0.15 M NaCl, pH = 8.1, and 0.05% sodium azide.
   - **Standard and Sample buffer**: the Assay Base buffer is supplemented to 0.1% Triton X-100 and then to 2 mg/mL Rabbit IgG and filtered. The IgG is more soluble with the Triton X-100 present. Filtered buffer is stable for 1 month at 4 °C. Twenty-five mL buffer is required for one assay plate.
   - **Assay buffer**: The Assay Base buffer is supplemented to 0.5% Triton X-100 and then to 2 mg/mL Rabbit IgG and filtered. Filtered buffer is stable for 1 month at 4 °C. Twenty mL buffer is required for one assay plate.
6. SNAP-25 Assay Buffers
   - **Standard and Sample buffer**: Thermo Scientific Blocker Casein in TBS supplemented to 0.1% Triton X-100. This buffer is not filtered due to turbidity of casein solution and is stable for 1 week at 4 °C. Twenty-five mL buffer is required for one assay plate.
   - **Assay buffer:** Casein in TBS supplemented to 0.5% Triton X-100 and then to 2 mg/mL mouse IgG. 20 mL of assay buffer is required for one assay plate and is prepared as follows: Combine 19 mL TBS Casein +1 mL 10% Triton X-100 solution and mix briefly. Transfer 4 mL to a 10 mL tube. Add 40 mg mouse IgG and mix on a rotisserie rotator for 20 min. Filter this solution through a 0.2 μM syringe filter, adding it back to the 16 mL TBS casein buffer previously set aside. This means of preparation is used as mouse IgG may not fully dissolve and Casein solution is turbid, therefore, large volumes will rapidly clog a 0.2 μm filter. Buffer is stable for 1 week at 4 °C.
7. VILIP-1 Assay Buffer: 0.01 M TRIS, 0.15 M NaCl, 0.1% Triton X-100 and 0.05% Sodium Azide, pH 8.1, to which is then added 1 mg/mL BSA and 2 mg/mL mouse IgG. Filtered buffer is stable for 1 month at 4 °C. Add 0.08 mL of 1 M Calcium Chloride ($CaCl_2$) solution to 40 mL buffer (final concentration of CaCl2 is 2 mM) at the time of the assay, as over time $CaCl_2$ will precipitate out of solution, causing it to appear cloudy. All directions regarding this buffer in the future assume that $CaCl_2$ has been added. Forty mL buffer is required for one assay plate.

10.2.7 Capture Antibody Magnetic Microparticle Reagent

1. *Biotinylation of capture antibodies:* Uses Thermo Scientific Sulfo NHS-LC-LC-biotin (PN 21338) and the protocol provided by the vendor, with the following modifications: 0.1 M carbonate buffer at pH 9 instead of PBS (*see* Subheading 10.6.3 for an explanation) is used for the labeling reaction, and two-fold molar excess of biotin is used per mole of IgG instead of 20 molar excess to avoid over-labeling antibodies. Amicon Ultra-4, 30,000 MW cut-off centrifugal filter units are used for the buffer exchange steps (4 mL final buffer volume for 1 mg of IgG, for each exchange step). PBS supplemented to 0.1% sodium azide is used for storage of the biotinylated antibody at 1 mg/mL, which is stable for 6 months stored at 4 °C. Refer to Subheading 10.6.3 for additional information about the biotin labeling protocol.
2. *Coupling biotinylated antibodies to microparticles:* Biotinylated antibodies are coupled to Dynabeads MyOne™ Streptavidin C1 magnetic microparticles for the rabbit Ng, VILIP-1, and SNAP-25 assays or Dynabeads MyOne™ Streptavidin T1 magnetic microparticles for the monoclonal Ng assay (*see* Subheading 10.2.7, **step 6**), using the protocols provided by Thermo Fisher with a few modifications. Borate-buffered saline [BBS] (*see* Subheading 10.2.4, **step 1**) supplemented to 0.1% Triton X-100 [BBST] containing no azide, is used for the coupling reaction. To remove the manufacturer's storage buffer (PBS, 0.1% Tween-20 and sodium azide), uncoated C1 microparticles are washed three times with BBST before coupling to biotinylated antibodies. For coupling, incubate the washed microparticles and biotinylated antibody for 1 h at room temperature on a Rotisserie rotator. After coupling, coated microparticles are washed 5 times with BBST and then resuspended at 10 mg/mL in BBST supplemented to 0.1% sodium azide. Coated microparticles are stored in 65 uL aliquots, a volume sufficient for one assay plate. Coated microparticles stored at 4 °C are stable for 6 months.
3. *Rabbit Ng Assay Capture Antibody "Microparticle" Reagent:* Biotinylated rabbit affinity purified p4793 antibody is coupled to MyOne C1 microparticles at a concentration of 6.25 μg IgG per mg of microparticles. Coated microparticles are prepared and stored as described in Subheading 10.2.7, **step 2**.
4. *SNAP-25 Assay Capture Antibody "Microparticle" Reagent:* Biotinylated monoclonal 6H07 antibody is coupled to MyOne C1 microparticles at a concentration of 12.5 μg IgG per mg microparticles. Coated microparticles are prepared and stored as described in Subheading 10.2.7, **step 2**.
5. *VILIP-1 Assay Capture Antibody "Microparticle" Reagent:* Biotinylated monoclonal 3A8 antibody is coupled to MyOne C1 microparticles at a concentration of 25 μg IgG per mg

microparticles. Coated microparticles are prepared and stored as described in Subheading 10.2.7, **step 2**.

6. *Monoclonal Ng Assay Capture Antibody "Microparticle" Reagent:* Biotinylated monoclonal 14D03 antibody is coupled to MyOne Streptavidin T1 (tosylactivated) Dynabeads and **NOT** C1 Dynabeads, as we observe that C1 (carboxylated) Dynabeads will aggregate during the 14D03 antibody-microparticle coupling step. Coupling of biotinylated antibody to T1 microparticles follows the Thermo Scientific protocol and uses the modifications described in Subheading 10.2.7, **step 2**) with the exception that uncoated T1 microparticles are washed a total of five (not three) times with BBST, to remove the manufacturer's storage buffer (PBS, 0.1% BSA and sodium azide) before coupling, which reduces background signal in the Monoclonal Ng assay. Coating concentration is 25 ug IgG per mg microparticles. Coated microparticles are stored as described in Subheading 10.2.7, **step 2**.

10.2.8 Labeled Detection Antibody Reagent Preparation

Invitrogen Alexa® Fluor 647 dye labeled "detection" antibody is prepared using Millipore Sigma's Erenna® SMC™ Detection Reagent Labeling Kit (03–0076-02) following manufacturer's protocol without modification. Labeled detection antibody is stored at 1 mg/mL in PBS (labeling kit buffer #3), supplemented to 0.1% sodium azide. Stability is 6 months at 4 °C. As this concentrated labeled detection antibody may form aggregates over time, and to improve its stability if there is frequent use, it is recommended that the labeled antibody be stored in 100–150 μL aliquots (each sufficient for 15–20 assay plates). Discard an aliquot after 10 uses. The optimal degree of labeling is ~2–4 molecules of dye per molecule of IgG, calculations for which are provided in the Subheading 10.6.4: Degree of labeling of Alexa® Fluor dye.

10.2.9 Quality Control Samples

Quality control CSF samples (QC) are human CSF at low, medium, and high concentrations. These sets of QC controls are specific to each assay and are necessary to track assay performance over time. For QC samples, we utilized excess human CSF from patients with hemolytic stroke being treated with a shunt. The CSF draining from the shunt is generally collected in a large vessel at room temperature, which is disposed of every 24 h. Samples from a vessel with an appropriate concentration are aliquoted into small tubes and frozen. QC CSF samples were obtained via a human studies protocol approved with a waiver of consent, by the institutional review board at Washington University (St. Louis, MO, USA), as the de-identified samples were all scheduled for disposal. We have no experience with commercial CSF controls, but they may also be suitable.

10.2.10 Assay Standardization

1. Our preferred method for quantitating synthetic or recombinant proteins used for standardization is amino acid analysis (AAA), as we observe that weighing (Ng synthetic protein), Bradford assay (SNAP-25) and A280 (VILIP-1) quantitation methods overestimated the concentration of these proteins up to two-fold compared to amino acid analysis. This section describes preparation of stock assay standards which are stored as "single use" aliquots at −80 °C in the assay-specific buffers described in Subheading 2.6. These stock standards are prepared at concentrations that are stable and allow a single dilution of the stock standard (or two dilutions for VILIP-1 stock standard) to create Standard 1 for the assay. Instructions for preparing assay-specific standard curves from these stock standards are provided later in Subheading 10.3.4.
2. *Monoclonal Ng Assay Standard:* Synthetic full length Ng from AAPPTec, Inc. [Louisville KY] is used as the standard. Synthetic Ng was weighed into PBS, pH 7.4 and then quantitated using AAA at 0.5 mg/mL. This is used to prepare a stock standard at 250 ng/mL in **Monoclonal Ng Assay buffer** (*see* Subheading 10.2.6, **step 4**) that is stored in 0.1 mL aliquots at −80 °C.
3. *Rabbit Ng Assay Standard:* Synthetic full length Ng from AAPPTec, Inc. is used as the assay standard. Synthetic Ng was weighed into PBS, pH 7.4, and then quantitated using AAA at 0.5 mg/mL. This is used to prepare a stock standard at 250 ng/mL in **Rabbit Ng Standard and Sample buffer** (*see* Subheading 10.2.6, **step 5**) that is stored in 0.1 mL aliquots at −80 °C.
4. *SNAP-25 Assay Standard*: Cell Sciences (Newburyport, MA) recombinant human SNAP-25-His (CSI15602), in 20 mM Tris-HCl, pH 7.5, 50 mM NaCl, 2 mM DTT, and 1 mM EDTA) is used as the assay standard and was quantitated using AAA at 0.69 mg/mL. This is used to prepare a 50 ng/mL stock standard in **SNAP-25 Standard and Sample buffer** (*see* Subheading 10.2.6, **step 6**) that is stored in 0.1 mL aliquots at −80 °C.
5. *VILIP-1 Assay Standard:* N-myristoylated His-tagged Human VILIP-1 monomer in PBS, pH 7.4, plus 0.05% sodium azide) from Proteos, Inc. (Kalamazoo, MI), is used as the assay standard, and was quantitated using AAA at 1.32 mg/mL [14]. This is used to prepare stock standard at 13.2 μg/mL in **VILIP-1 Assay buffer** (*see* Subheading 10.2.6, **step 7**) that is stored in 0.1 mL aliquots at −80 °C.

10.3 Assay Preparation

10.3.1 Technical Guidelines

1. Clean bench surface and pipettes with 70% ethanol before use. Keep bench surface and supplies free of dust which can interfere in the assay causing false high fluorescent signal and poor reproducibility.
2. Allow all buffers to warm to room temperature prior to use.
3. Sterilize Automated Microplate Washer daily before use, with 70% ethanol. Prime the washer liquid lines with 70% ethanol and let sit for 30 min. Rinse ethanol from lines with three changes of purified water, followed by three changes of wash buffer to remove bubbles from the lines.
4. Two different Axygen 96 V-well plates are used for each assay. In addition, a Nunc 384-well polypropylene plate is used for reading eluted Alexa® Fluor-labeled antibodies and can include up to four 96-well assays. The three plates are:
   - The V-well "assay plate" (referred to as **Plate 1**) is used for the assay capture and detection steps.
   - A separate V-well "transfer plate" (referred to as **Plate 2**) is used for the elution of labeled detection antibody. The manual transfer of magnetic microparticles from the V-well assay plate (**Plate 1**) to a clean V-well plate ("Transfer plate" or **Plate 2**), following the Post-Detection Wash step (Subheading 10.4.4), is essential for reducing possible background signal, as the V-well plates are not blocked. This step (described in Subheading 10.4.5) requires careful mixing and pipetting to transfer the entire contents of wells from one plate to the other. In our experience, detection antibodies eluted from microparticles without this transfer step, show highly variable signal and poor replicate CVs.
   - The 384-well Nunc polypropylene microplate used for reading eluted detection antibodies in the Erenna is referred to as **Plate 3**.

10.3.2 CSF Sample Preparation

1. We analyze CSF samples with aliquots frozen when the CSF is obtained. Avoid multiple freeze thaw cycles of CSF. Optimally samples should be assayed on the first thaw. When assayed on the second thaw, we observe some decrease (up to 10%) in the VILIP-1 and Ng assay results which can vary among samples (unpublished data), and a further decrease of ~10% after a second thaw. Even greater decrease is observed in SNAP-25 concentrations, which appear to be sensitive to extended thaw times as well as number of thaws. Thaw CSF samples and Quality Control (QC) samples just prior to use at 4 °C in a refrigerator or in wet ice. A 0.5 mL aliquot typically requires one hour to thaw at 4 °C.
2. Centrifuge CSF at 12,000 ×*g* for 3 min at 4 °C prior to use to separate particulate matter. Place CSF samples in a tube rack set

in crushed ice. CSF stored on top of ice may not be cold enough.

3. CSF samples may be assayed in duplicate or triplicate. Refer to Table 12.5 for minimum volume per well for each of the immunoassays.

10.3.3 Capture Antibody Reagent Preparation

1. For all of the analytes, capture antibody-coated microparticles (at 10 mg/mL) are stored in 65 μL aliquots (as described in Subheading 10.2.7, **step 2**: Capture Antibody Reagent Preparation).
2. For all of the analytes, the working capture antibody reagent is made by diluting 65 μL microparticles (at 10 mg/mL) to 13.0 mL (which is a concentration of 50 μg microparticles/mL) in a 15 mL tube, in the respective **Assay buffer** for each analyte. 0.1 mL microparticles are added per well: 0.1 mL × 50 μg/mL microparticles = 5 μg microparticles/well.
3. The Capture Antibody-microparticle reagent is mixed well on a rotisserie rotator for 30 min before addition of standards and samples to enable blocking of unbound sites on the antibody-coated microparticle surface.

10.3.4 Standard Curve Preparation

1. Thaw a "single use" stock standard vial (described for each assay in Subheading 10.2.10) just prior to use. Invert the vial several times to mix the contents and briefly spin to collect contents in bottom of vial, before opening. Standards are not vortexed to avoid possible sticking to the vial.
2. All assays use an 11-point standard curve, assayed in triplicate, loaded as vertical replicates across the top of the plate, columns 1 to 11. VILIP-1 and SNAP-25 assay standard curves use serial dilutions, while the rabbit and monoclonal Ng assay standards use parallel dilutions. This will be explained for each assay. Refer to Subheading 10.2.6 for directions for making buffers used for standard curve preparation for each of the assays. Prepare standards at room temperature and ensure they are mixed thoroughly at every step.
3. For improved precision, we recommend that Standard 1 for all assays (and the 132 ng/mL intermediate used to prepare Standard 1 for the VILIP-1 assay) be prepared twice independently and pooled for greater accuracy.
4. *Monoclonal Ng Assay Standard Curve Preparation.* Label polypropylene tubes 1 through 11. The curve is prepared using **Monoclonal Ng Assay buffer** (*see* Subheading 10.2.6, **step 4**). Use a new pipette tip for every transfer.
   - To prepare Standard 1, an aliquot of the 250 ng/mL Ng stock standard is thawed in ice immediately before use,

Table 12.3
Schematic for standard curve preparation for the monoclonal Ng assay (*see* Subheading 10.3.4, step 4)

Tube no	1	2	3	4	5	6	7	8	9	10	11
Buffer vol (μL)		2100	333	667	900	933.3	966.7	978	989	995	500
Tube 1 (3000 pg/mL) vol (μL)		900									
Tube 2 (900 pg/mL) vol (μL)			667	333	100	66.6	33.3	22.2	11.1	5.55	
Ng (pg/mL)	3000	900	600	300	90	60	30	20	10	5	0

Prepare Standard 1 (3000 pg/mL) as directed above (Subheading 10.3.4, *step 4) using 250 ng/mL stock standard "single use" aliquot. Standard 2 combines 900 μL of Standard 1 and 2100 μL buffer. Standards 3 through 10 are parallel dilutions of Standard 2 (900 pg/mL), using increasing volumes of buffer and decreasing volumes of the 900 pg/mL stock.*

Table 12.4
Schematic for standard curve preparation for the Rabbit Ng assay (*see* Subheading 10.3.4, step 5)

Tube no	1	2	3	4	5	6	7	8	9	10	11
Buffer vol (μL)		500	750	1800	500	750	900	950	975	990	500
Tube 1 (800 pg/mL) vol (μL)		500	250	200							
Tube 2 (80 pg/mL) vol (μL)					500	250	100	50	25	10	
Ng (pg/mL)	800	400	200	80	40	20	8	4	2	0.8	0

Prepare Standard 1 (800 pg/mL) as directed above (Subheading 10.3.4, *step 5) using 250 ng/mL stock standard "single use" aliquot. Standards 2 through 4 are parallel dilutions of Standard 1, using increasing volumes of buffer and decreasing volumes of the 800 pg/mL stock. Standards 5 through 10 are parallel dilutions of Standard 4 (80 pg/mL), using increasing volumes of buffer and decreasing volumes of the 80 pg/mL stock.*

mixed gently, and diluted to 3000 pg/mL by diluting 24 μL of the 250 ng/mL stock to 3.0 mL with Assay buffer.

- Standard 2 at 900 pg/mL is prepared by combining 0.9 mL of Standard 1 plus 2.1 mL Assay buffer in tube 2.
- Standards 3 through 10 are "parallel" dilutions of Standard 2 (900 pg/mL), combining decreasing volumes of Standard 2 and increasing volumes of assay buffer. In tube 3, combine 666 μL Std 2 + 333 μL assay buffer (600 pg/mL) and mix. Continue as depicted in Table 12.3. Tube 11 is buffer only.

5. *Rabbit Ng assay Standard Curve Preparation.* Label polypropylene tubes 1 through 11. Use a new pipette tip for every transfer. The curve is prepared using **Rabbit Ng Standard and Sample buffer** (*see* Subheading 10.2.6, **step 5**).
   - To prepare Standard 1 (800 pg/mL), an aliquot of the 250 ng/mL Ng stock standard is thawed in ice immediately before use, mixed gently, and diluted to 800 pg/mL by diluting 16 μL of the 250 ng/mL stock to 5.0 mL with standard buffer in tube 1.

- Standards 2 (400 pg/mL), 3 (200 pg/mL) and 4 (80 pg/mL) are parallel dilutions of Standard 1 (800 pg/mL), combining decreasing volumes of Standard 1 and increasing volumes of standard buffer. In tube 2, combine 500 μL Std 1 + 500 μL standard buffer (400 pg/mL) and mix. Continue as depicted in Table 12.4 preparing Standards 3 and 4.
- Standards 5 through 10 are parallel dilutions of Standard 4 (80 pg/mL), combining decreasing volumes of Standard 4 and increasing volumes of buffer. In tube 5, combine 500 μL of Standard 4 and 500 μL buffer. Continue as depicted in Table 12.4.

6. *SNAP-25 Assay Standard Curve Preparation.* Label tubes 1 through 11. Use a new pipette tip for every transfer. The curve is prepared using **SNAP-25 Standard and Sample buffer** (Subheading 10.2.6, **step 6**).
   - To prepare Standard 1(90 pg/mL), an aliquot of the 50 ng/mL SNAP-25 stock standard is thawed on ice immediately before use, mixed gently, and diluted to 90 pg/mL by diluting 9 uL of the 50 ng/ml stock standard to 5.0 mL with standard buffer in tube 1.
   - Standards 2 (30 pg/mL) and 3 (10 pg/mL) are three-fold serial dilutions of Standard 1. Add 1.0 mL of buffer to tubes labeled 2 and 3. Transfer 0.50 mL of Standard 1 to tube 2, mix thoroughly. Transfer 0.5 mL of tube 2 to tube 3.
   - Standards 4 through 10 are two-fold serial dilutions of Standard 3 (10 pg/mL). Add 0.5 mL of buffer to tubes labeled 4 through 11. Transfer 0.5 mL from tube 3 to tube 4, mix thoroughly. Continue serial dilutions as depicted in Fig. 12.4. Tube 11 (0 pg/mL) is buffer only.

Figure 4: Schematic for Standard Curve Preparation for the SNAP-25 Assay

Tube #	1	2	3	4	5	6	7	8	9	10	11
Transfer vol (uL)		500	500	500	500	500	500	500	500	500	
Buffer vol (uL)		1,000	1,000	500	500	500	500	500	500	500	500
SNAP-25 (pg/mL)	90	30	10	5	2.5	1.25	0.625	0.313	0.156	0.078	0

Standard 1 (90 pg/mL) is prepared as directed above (9 uL of 50 ng/mL stock standard + 4.991 mL). Standards 2 and 3 are 3-fold serial dilutions of Standard 1. Standards 4 through 10 are 2-fold serial dilutions of Standard 3. Standard 11 is buffer only.

Fig. 12.4 Schematic for standard curve preparation for the SNAP-25 assay (Subheading 10.3.4, **step 6**)

Figure 5: Schematic for Standard Curve Preparation for the VILIP-1 Assay											
Tube #	1	2	3	4	5	6	7	8	9	10	11
Transfer vol (uL)		200	200	200	200	200	200	200	200	200	
Buffer vol (uL)	1,970	200	200	200	200	200	200	200	200	200	200
Intermediate 132 ng/mL (uL)	30.3										
VILIP-1 (pg/mL)	2,000	1,000	500	250	125	62.5	31.3	15.6	7.8	3.9	0

As directed above, use the 13.2 ug/mL stock standard to prepare a 132 ng/mL intermediate, by combining 10 uL of stock standard + 0.99 mL buffer. Standard 1 (2,000 pg/mL) is prepared by diluting 30.3 uL of the intermediate to 2.0 mL. Standards 2 and 10 are 2-fold serial dilutions of Standard 1. Standard 11 is buffer only.

Fig. 12.5 Schematic for standard curve preparation for the VILIP-1 assay (Subheading 10.3.4, **step 7**)

7. *VILIP-1 Assay Standard Curve Preparation*. Label a polypropylene tube "132 ng/mL" and also tubes 1 through 11. Use a new pipette tip for every transfer. The curve is prepared using **"VILIP-1 Assay buffer"** (*see* Subheading 10.2.6, **step** 7).
   - An aliquot of the "single use" 13.2 μg/mL "VILIP-1 stock standard is thawed on ice immediately before use, mixed gently, and diluted 1:100 by diluting 10 μL of the 13.2 μg/mL stock standard to 1.0 mL with assay buffer, yielding a 132 ng/mL intermediate which is then used to prepare Standard 1 (2000 pg/mL). For added accuracy, prepare the 132 ng/mL intermediate twice and pool the two, before using to prepare Standard 1.
   - Standard 1, at 2000 pg/mL, is prepared by diluting 30.3 μL of the 132 ng/mL intermediate to 2.0 mL with assay buffer.
   - Standards 2 through 10 are two-fold serial dilutions of the Standard 1 (2000 pg/mL), performed in polypropylene tubes. Add 0.2 mL of buffer to tubes labeled 2 through 11. Transfer 0.2 mL of Standard 1 to tube #2, mix thoroughly. Continue serial dilutions as depicted in Fig. 12.5. Tube 11 is buffer only.

10.4 Assay Procedure (All Assays)

10.4.1 Capture Step Setup Protocols

1. It is recommended that for the capture step only, an orbital plate shaker set at ~525 rpm that is designed to accelerate slowly to the selected speed, be used. The 200 μL well volume is less likely to splash the adhesive sealing film plate cover when acceleration is gradual.

Table 12.5
Assay capture step component volumes for assays

Assay	Sample or standard	CSF dilution	Component volumes per well				Total per well
			Sample dilution buffer	MPs	CSF sample	Standard	
mAb Ng	Standard		~	100 μL	~	100 μL	200 μL
	CSF sample	1:10	90 μL	100 μL	10 μL	~	
Ng, rabbit	Standard		~	100 μL	~	100 μL	200 μL
	CSF sample	1:20	95 μL	100 μL	5 μL	~	
SNAP-25	Standard		~	100 μL	~	100 μL	200 μL
	CSF sample	1:4	75 μL	100 μL	25 μL	~	
VILIP-1	Standard		75 μL	100 μL	~	25 μL	200 μL
	CSF sample	Neat	75 μL	100 μL	25 μL	~	

Note: Test CSF and QC CSF samples use the same dilution, in each of the assays
For all assays, add components in the order: sample dilution buffer, magnetic microparticles (MPs), CSF, then Standards. For the VILIP-1 assay add 75 μL of buffer to ***all*** *wells, then microparticles, then 25 μL CSF or Standards*

2. Buffer used to dilute CSF samples is added first to plate wells, then microparticles, then samples and standards, as we observe slight suppression in Ng and SNAP-25 concentrations if CSF is first diluted in buffer then added to microparticles.
3. Per well volumes of assay components added directly to wells of 96 V-plate "assay" plate (referred to as **Plate 1**) are provided in Table 12.5. Components are listed in the order they should be added. Note: The working capture microparticle reagent prepared in Subheading 10.3.3 is abbreviated as MPs in Table 12.5. In each assay, the QC CSF samples and CSF test samples use the same dilution.
4. Seal **Plate 1** with an adhesive sealing film plate cover. Set on orbital plate shaker and mix at ~525 rpm 2 h.
5. At the end of the 2-h capture step, carefully remove the plate sealing film.

10.4.2 Post-Capture Wash Protocol

1. If using a Tecan Hydroflex™ Automated Plate Washer, run a wash protocol that adds 300 μL/well wash buffer, followed by aspiration, two times.
2. If using a manual Post-Capture Wash Protocol proceed as follows:
   (a) Place V-well assay plate (**Plate 1**) onto magnet plate (DynaMag™ -96 Magnet). Wait 2 min and confirm all microparticles are pulled down into a pellet.
   (b) Aspirate the liquid from wells (microparticles will remain visible in well)

(c) Add 300 μL Wash Buffer. Wait 1 min.

(d) Aspirate the wash buffer from wells.

(e) Repeat **steps 10.4.2.2c** and **10.4.2.2d** one more time for a total of two washes.

10.4.3 Detection Step Protocols

1. Prepare detection antibody reagent, as described below, 10 min before the end of the capture step. Protect from light before use.
2. As concentrated Alexa® Fluor Detection antibody reagent at 1 mg/mL (described in Subheading 10.2.8) may form aggregates during storage, prepare a 1:100 dilution (10,000 ng/mL), by diluting 5 μL labeled Detection antibody reagent to 0.5 mL with assay buffer in an Ultrafree-MC 0.2 um centrifugal filter. Centrifuge the UltraFree filter for 30 s × 11,000 g. Remove the filter cup, recap the vial, and gently mix by inverting the vial several times. For each of the analytes, use this 10,000 ng/mL detection antibody reagent to prepare 4 mL of working detection antibody reagent, using concentrations specific to each assay. Instructions for each assay are as follows:
3. *Monoclonal Ng Assay Working Detection Antibody Reagent.* The detection antibody is used at 100 ng/mL. 4 mL of the working reagent at 100 ng/mL is prepared by diluting 40 μL of the filtered 10,000 ng/mL detection antibody reagent to 4.0 mL with **Monoclonal Ng Assay Buffer** (Subheading 10.2.6, **step 4**). Mix gently by inverting several times.
4. *Rabbit Ng Assay Working Detection Antibody Reagent.* This Detection antibody is used at 25 ng/mL. 4 mL working reagent at 25 ng/mL is prepared by diluting 9.6 μL of the filtered 10,000 ng/mL detection antibody reagent to 4.0 mL with **Rabbit Ng assay buffer** (Subheading 10.2.6, **step 5**). Mix gently by inverting several times.
5. *SNAP-25 Assay Working Detection Antibody Reagent.*
   - For the SNAP-25 assay only, it is necessary to filter 2–3 mL of **SNAP-25 Assay buffer** (Subheading 10.2.6, **step 6**) through a 0.2 μm syringe filter prior to preparation of the 1:100 dilution of detection antibody stock in the 0.5 mL Ultrafree Centrifugal filter. Unfiltered assay buffer (which is turbid due to the casein component) will rapidly clog the centrifugal filter resulting in incomplete filtering of the antibody reagent.
   - The working detection antibody is used at 500 ng/mL for the SNAP-25 assay. 4 mL of working reagent at 500 ng/mL is prepared by diluting 200 μL of the filtered 10,000 ng/mL detection antibody reagent to 4.0 mL with **SNAP-25 assay**

buffer (Subheading 10.2.6, **step 6**). Mix gently by inverting several times.

6. *VILIP-1 Assay Working Detection Antibody Reagent.* Detection antibody is used at 100 ng/mL. 4 mL of working reagent at 100 ng/mL is prepared by diluting 40 μL of the filtered 10,000 ng/mL detection antibody reagent to 4.0 mL with **VILIP-1 assay buffer** (Subheading 10.2.6, **step** 7). Mix gently by inverting several times.
7. For all assays, immediately following the Post-Capture wash step, add 20 μL of working detection reagent to each well with an 8-channel pipette. Seal the plate with an adhesive sealing film plate cover and mix at ~800 rpm for 1 h. Protect from light.
8. A Stat Fax® 2200 Microplate Incubator/Shaker (or equivalent) is recommended for detection and elution steps, as tinted cover provides protection from light. Use speed setting ~800 rpm and ambient temperature setting.
9. At the end of the 1-h detection step, carefully remove the plate sealing film.

10.4.4 Post-Detection Wash Protocol

1. If using a Tecan Hydroflex™ Automated Plate Washer, run a wash protocol that adds 250 μL/well of wash buffer followed by aspiration, three times, and then leaves the fourth wash volume in the wells. Before running the wash protocol, and to ensure that magnetic microparticles will be well separated from the aspiration pins, add 100 μL wash buffer to each well using an 8-channel pipet.
2. If using a manual Post-Detection Wash Protocol proceed as follows:
   (a) Place V-well assay plate (**Plate 1**) onto DynaMag™-96 magnet. Add 100 uL Wash Buffer to each well using an 8-channel pipet. Wait 2 min to allow microparticles to form a pellet. Aspirate the fluid from the wells and discard. Change tips.
   (b) Add 250 μL Wash Buffer to each well. Wait 1 min. Aspirate buffer from each well and discard. Change tips.
   (c) Repeat **step 10.4.4.2b** two more times.
   (d) Add 250 μL of Wash Buffer to each well (for a total of 4 washes).

10.4.5 Manual Plate 1 to Plate 2 Transfer

1. The manual transfer of microparticles from the V-well **Plate 1** to a clean V-well plate ("Transfer plate" or **Plate 2**), following the Post-Detection Wash step, is essential for reducing background signal (*see* also Subheading 10.3.1, **step 4**). This step requires careful mixing to resuspend microparticles and pipetting to transfer entire contents of wells from one plate to the other.

2. This protocol describes transfer of microparticles from **Plate 1** to **Plate 2**, one row at a time, using a 12-channel pipette.
3. Following Post-Detection Wash, remove V-well **Plate 1** from magnet. Set **Plate 1** on orbital plate shaker set to ~575 rpm and mix for 2 min to resuspend microparticles.
4. Label a new V-well plate (defined as **Plate 2**) and set it on DynaMag magnet.
5. Set 12-channel pipette volume to 125 μl. Put clean tips from row 1 of a multichannel pipette tip box (96 tips) onto pipette.
6. Remove **Plate 1** from plate shaker and set it on a sheet of white paper on the lab bench (to better visualize the microparticles). Use the same orientation for **Plate 1** on bench as **Plate 2** on magnet.
7. In the first row of **Plate 1**, gently pipette up and down 5× to fully resuspend microparticles.
8. Transfer *entire* Row A well contents (125 μL × 2) of **Plate 1** to Row A of **Plate 2**. Work slowly and gently to minimize bubbles.
9. Change pipette tips. Mix 5× and transfer entire Row B well contents (125 μL × 2) of **Plate 1** to Row B of **Plate 2**. Repeat these steps for all rows.
10. If any microparticles remain in **Plate 1** upon visual inspection, add 125 μL of wash buffer from the corresponding well of **Plate 2**. Pipette mix and transfer microparticles from **Plate 1** to corresponding well of **Plate 2**.

10.4.6 Final Aspiration

1. To improve assay replicate precision, bubbles must be removed from wash buffer surface in transfer **Plate 2**, prior to final wash buffer aspiration step and addition of elution buffer. To remove bubbles, gently blow air over the surface of the plate using a 1 mL single channel pipette fitted with a clean tip. Alternatively, if the lab bench is equipped with compressed air, attach clean tubing with a filter on the end to trap dust or moisture and adjust to a gentle flow several inches above the plate surface, to efficiently remove bubbles.
2. If using a Tecan Hydroflex™ Automated Plate Washer, run an "aspiration only" protocol to remove wash buffer from all wells.
3. If using a manual Final Aspiration Protocol, set **Plate 2** on Dynamag® magnet and wait 2 min. Working one row at a time, aspirate wash buffer using an 8-channel pipet from wells and discard.

10.4.7 Detection Antibody Elution Protocol

1. The 384-well Nunc polypropylene microplate used for reading eluted detection antibodies in the Erenna (referred to as **Plate 3**) is rinsed with purified water before use. This will remove dust or debris that may interfere with reading samples in the Erenna. Pipette 125 μL purified water per well. After discarding water from the wells, place plate upside down on paper towel in plate carrier and then centrifuge for 1 min at 1900 ×*g*. Keep plate covered until ready to use.
2. The 384-well **Plate 3** contains four quadrants of 96 wells, 24 columns × 16 rows, allowing for up to four 96-well plate assays (or experiments) to be defined within a "test" using Erenna SGX Link software. To minimize errors and assist in the transfer of eluted detection antibodies from each 96-well plate to the correct 384-well plate quadrant, underline the odd numbers of the 24 columns. Quadrant 1 wells are always in odd numbered columns and odd lettered rows (wells A1, A3, A5...A23, C1, C3, C5...C23 etc), quadrant 2 wells in even numbered columns and odd lettered rows (A2, A4, A6...A24, etc), quadrant 3 wells in odd numbered columns and even lettered rows (B1, B3, B5...B23, D1, D3, D5...D23, etc.), quadrant 4 wells in even numbered columns and even lettered rows (B2, B4, B6...B24, etc.). Use an 8-channel pipet to transfer one 96-well plate column at a time, positioning the top tip relative to the underlined column number to locate the correct quadrant.
3. Elution steps use Elution Buffer B (glycine) and Buffer C (1 M TRIS for neutralization) from Millipore Sigma (*see* Subheading 10.2.3, **step 1**) as well as 4titude Pierce Seal Plate Foil covers (*see* Subheading 10.2.2, **step 1**) for heat sealing a pierceable foil cover to Nunc 384-well **Plate 3**.
4. Immediately following "Final Aspiration" (*see* Subheading 10.4.6) remove V-well **Plate 2** from the magnet and add 21 μL Elution Buffer B per well with an 8-channel pipette.
5. Seal **Plate 2** with an adhesive sealing film plate cover and mix for 10 min on StatFax plate shaker at ~800 rpm.
6. During the 10-min detection antibody elution step, add 4 μL per well of Buffer C to **Plate 3** and seal with adhesive sealing film.
7. Prior to transfer of eluted detection antibodies from **Plate 2** to **Plate 3**, turn on FluidX X-Seal Thermal Sealer (or equivalent device). Set Thermal Sealer to a sealing temperature of 170 °C with 2.5 s sealing time, when used with 4titude Pierce Seal Plate Foil covers (per manufacturer's instructions).

8. Carefully remove sealing film from **Plate 2** and set plate on DynaMag magnet. Allow microparticles to form a tight pellet for 2 min.
9. Set the 8-channel pipette to 20 μL. Place 8 tips on the pipette.
10. The transfer of eluted detection antibodies from the **Plate 2** to the **Plate 3** requires care. To avoid creating bubbles (the prime cause of false high signal TP fliers) a single, slow, smooth aspiration of eluate from wells of **Plate 2** and single slow dispense into **Plate 3** is recommended. Being able to perform a single, slow aspiration (while varying the tilt of the tips slightly to avoid air-locking the tips on the bottom on the well) is very desirable. After sealing the plate with a 4titude Pierce Seal Plate Foil cover, centrifuging the sealed plate for 4 min at high speed (~1900 g) will remove most bubbles.
11. Transfer the eluate from **Plate 2** to **Plate 3**. Work slowly and carefully (see also instructions in Subheading 10.4.7, **step 2**), transferring one plate column at a time, changing tips between each column.
12. Seal **Plate 3** with a 4titude Pierce Seal Plate Foil cover. To remove bubbles created during eluate transfer, place sealed **Plate 3** in plate carrier (right side up), and centrifuge for 4 min × 1900 ×*g*.
13. Load sealed **Plate 3** onto the Erenna® for reading.

10.5 Data Analysis

10.5.1 Calculation of Results

1. During the reading on the Erenna, labeled detection antibodies are drawn into the instrument's capillary line, then passed through a laser which excites the Alexa® Fluor dye to emit photons. Signal with peak intensity above background fluorescence, using a threshold of at least eight times background, corresponds to a single molecule or detected event.
2. Curve fitting is done by weighted regression of three signal types: detected events (DE) (i.e., single molecules), event photons (EP), the sum of photons associated only with detected events, and total photons (TP), the sum of all photons from events and background, using a proprietary algorithm (SMDCurve Fit, Singulex Software SGXLink). The algorithm generates a curve using a 5 parameter logistic equation for each of these parameters. As with typical ELISAs, each curve tends to be flat at the high and low ends and have a linear portion in the middle. Depending on the relative concentration of the sample, each curve contributes a weighted percentage toward the combined algorithm. For very low signal samples, weighting favors the linear region of the DE curve. For mid and high range signal samples, weighting favors the linear regions of the EP and TP curves, respectively.

Table 12.6
Performance characteristics for Ladenson Lab immunoassays

Assay, (N assay days)	Number of QC results	QC High CSF			QC Mid CSF			QC Low CSF			
		Mean [pg/mL]	% CVRw	% CVr	Mean [pg/mL]	% CVRw	% CVr	Mean [pg/mL]	% CVRw	% CVr	LLoQ
Monoclonal Ng [26]	45	**1540**	9.4	6.1	**458**	15.0	9.9	**150**	20.9	17.9	5
Rabbit Ng (glycine) [31]	58	**5510**	7.8	5.2	**2490**	11.8	8.5	**840**	14.4	6.7	0.8
SNAP-25 [53]	132	**18.8**	12.5	7.8	**10.7**	13.2	8.2	**4.1**	11.7	8.9	0.1
VILIP-1 [53]	142	**548**	3.2	2.7	**126**	8.1	8.0	**45**	8.1	6.2	3.9

Performance parameters include inter-assay precision (between run) CVRw and intra-assay (within run) CVr, of quality control samples, calculated according to Andreasson [34] supplemental data. Lower limit of quantification (LLoQ) is defined as the lowest standard for which replicate CV is <20%, and observed/expected concentration × 100 is within the range of 80–120%. Note that the Ng assays use the same sources of high, mid, and low Ng CSF

10.5.2 Acceptance of Results

1. A number of factors can indicate a problem with an entire assay or just one or more samples. These can be identified via the measured concentrations for QC samples (see Table 12.6 for our QC data). Such problems include poor precision among replicates, spuriously high TP signal caused by bubbles or dust in the environment, or by outlier replicate concentrations in the standard curve. These are discussed more fully in Sect. 10.6.1 Evaluating Assay Failure.
2. Our quality control includes analysis of three different quality control samples (*see* Subheading 10.2.9) measured on each assay plate. QC mean and tolerance limits are best established by computing the average of at least 10 measured concentrations collected over at least 4 analytical runs prior to assaying study samples, if possible. Tolerance limits are defined at ±2 standard deviations (2SD) and ±3 standard deviations (3SD). Plates with two or more QC sample concentrations greater than two SDs from the mean (Westgard rule 2_{2s}) or one or more concentrations greater than three SDs from the mean (Westgard rule 1_{3s}) are re-analyzed [35]. Any individual sample with a coefficient of variation (% CV) greater than 25% between duplicates is re-analyzed. Any sample with a concentration exceeding the highest standard should be diluted and re-analyzed, or reported as > upper limit of quantification (ULoQ), and sample with a concentration lower than the highest standard should be reported as <LLoQ.

3. Precision performance characteristics for the monoclonal Ng, rabbit Ng, SNAP-25 and VILIP-1 assays are shown in Table 12.6. These characteristics include inter- and intra-assay CVs (defined as the repeatability of measurements of QC samples between and within assays, calculated according to Andreasson et al. supplemental data [34], and lower limit of quantification (LLoQ, defined as the lowest standard for which replicate CV is <20%, and observed [i.e., back-interpolated] concentration divided by the expected concentration × 100 is within the range of 80–120%). Note that the Ng assays use the same source of high, mid, or low Ng human CSF.

10.6 Notes

10.6.1 Evaluating Assay Failure

Assay failure may be reduced by regular maintenance, calibration, and performance evaluation of all instruments and equipment. Bubbles in eluted detection antibody transferred to the Nunc 384-well reading **Plate 3** will interfere in instrument reading causing occasionally high signal for Total Photons, sometimes called "fliers."

Operator errors can occur with any assay step, with varying consequences. Failure to correctly add an essential component (i.e., $CaCl_2$) to a buffer or to accurately prepare sample and standard dilutions can alter QC and sample concentrations. As the assays employ ~20 uL reagent volumes (i.e., the volume of working detection antibody and elution buffer), failure to completely aspirate wash buffer at the end of preceding wash or "final aspiration" steps dilutes the small reagent volume and leads to imprecision. Other factors that contribute to poor precision include incomplete mixing of reagents, working in a dusty environment (e.g., a lab bench covered with lab mat where dust cannot be seen), failing to filter a buffer that contains mouse or rabbit IgG which dissolves incompletely, bumping a plate, not removing plate sealing film with care, and forgetting to mix the plate to keep the magnetic microparticles suspended. Inter-assay CVs benefit from a consistent routine, with similar timing for the assay steps. Assay reproducibility is improved by routines that minimize the possibility of reagent and sample swaps or contamination e.g. Label ALL tubes and containers utilized. Use the same layout for the samples in the tube rack as the positions in the 96-well plate. Check order of samples in the tube rack against the plate map after the centrifugation step. Devise routines that enable you to discover errors that may easily occur. To assist an understanding of assay problems, keep detailed reagent lot documentation and record lot numbers of reagents used for each assay.

10.6.2 Adverse Interactions

EDTA interferes with VILIP-1 antibody–antigen binding and is excluded from VILIP-1 assay buffers. Blood samples containing EDTA can be accurately measured but by special techniques that convert EDTA plasma to heparinized plasma (unpublished data) that are beyond the scope of this chapter. EDTA may also affect Ng antibody–antibody binding. Bovine serum albumin (BSA) causes high background signal in the monoclonal and rabbit Ng assays.

10.6.3 Biotin Labeling of Capture Antibodies

As described in Subheading 10.2.7, **step 1**, biotinylation of capture antibodies uses Thermo Scientific Sulfo NHS-LC-LC-biotin (PN 2133843) per the protocol provided by the vendor, but with several modifications. A more detailed explanation is provided here. 0.1 M carbonate buffer at pH 9, instead of PBS, is used for the labeling reaction. The coupling of biotin to amine groups prefers an alkaline pH, and carbonate has better buffering capacity at higher pHs than phosphate. Tris is avoided as it can interfere with binding to amines. Carbonate has been used by others as well [36]. We suggest starting with a two-fold molar excess of biotin per mole of IgG to avoid over-labeling antibodies which may interfere with their ability to bind to antigen [37, 38]. Optimal degree of labeling is 1–2 molecules of biotin per molecule of IgG that can be determined using Thermo Scientific Pierce™ Biotin Quantitation Kit (28005), and that should be determined before addition of sodium azide to the biotinylated antibody.

Two Amicon Ultra-4, 30,000 MWCO centrifugal filter units are used for buffer exchange steps before and after labeling. Briefly, the protocol for labeling 1.0 mg of IgG is as follows: (1) Perform a buffer exchange into carbonate buffer using an Amicon filter, with 1 mg of IgG in 4 mL final buffer volume; (2) Centrifuge the Amicon filter tube for 10 min × 3900 ×*g* which will reduce the volume of IgG to ~0.25 mL within the filter; (3) Discard the flow-through buffer; (4) Perform a second buffer exchange adding carbonate buffer to a final volume of 4 mL and then centrifuge; (5) Transfer the remaining volume of IgG (~0.25 mL) to a polypropylene capped vial; (6) Determine the antibody concentration by reading the absorbance at A280 and adjust concentration to 1.0 mg/mL (A280 = 1.4); (7) Prepare 2.5 mM biotin solution by dissolving 4 mg biotin to 2.4 mL; (8) Immediately add 6 μL biotin per mg IgG; (9) Mix briefly and leave at ambient temperature for 1 h; (10) Neutralize unreacted biotin present in IgG sample by adding five times the biotin volume (i.e., 6 μL × 5) of 1 M TRIS buffer at pH 8; (11) Perform four exchanges into PBS using a second Amicon filter; (12) Determine the antibody concentration and adjust concentration to 1.0 mg/mL; (13) Add 0.1% sodium azide to the labeled antibody solution; and (14) sterile filter it using an Ultrafree centrifugal filter (*see* Subheading 10.2.2, **step 3d**). Labeled antibody is stable at 4 °C for 6 months.

10.6.4 Degree of Labeling for Alexa® Fluor Dye

Degree of labeling of molecules of dye per molecule of IgG is computed as instructed in Thermo Fisher Scientific's manual for Molecular Probes® "Alexa® Fluor" Antibody Labeling Kit.

Calculations employ molar extinction coefficients for the dye and IgG and measured absorbance readings at 280 and 650 nM. As noted in Subheading 10.2.8 and also originally recommended by Singulex Technical Service, optimal degree of labeling is 2–4 molecules of dye per molecule of IgG. In our experience, over-labeled antibody also causes higher background signal and contamination of the instrument capillary system.

10.6.5 Strategies for Improving Precision

1. As noted in Subheading 10.3.4, **step 3**, for improved precision, we recommend that Standard 1 for all assays (and the 132 ng/mL intermediate used to prepare Standard 1 for the VILIP-1 assay) be prepared twice independently and pooled for greater accuracy.
2. Subheading 10.4.1 provides directions for addition of assay components (sample dilution buffer, microparticles, QC and test samples and standards) directly to plate wells. However, as CSF sample volumes are quite small (5–25 μL per well), we suggest the following method that allows more accuracy. For this approach, a single preparation of each standard or diluted CSF sample is prepared for analysis in 8-tube strips of MicroTubes™ (*see* Subheading 10.2.2, **step 3b**). After mixing well, the contents of these tubes are split into plate wells for the binding/elution/counting steps and the results averaged. For this method, volumes in the MicroTubes™ are typically 2 (if duplicate assays) or 3 (if triplicate) times per well volume of each component (microparticles, buffer, standard or sample, as shown in Table 12.5), plus 10% extra volume to facilitate transfer. MicroTubes™ work well for combining components, as the tubes are spaced so that 8-channel pipette tips fit into them, to mix and transfer tube contents to the 96-well V-well plate. Before addition to the assay plate, mix tube contents by aspirating and pipetting 10–20 times using an 8-channel pipette set to 200 μL. This approach is especially helpful for the rabbit Ng assay which requires addition of 5 μL CSF per well.

10.6.6 Assay Throughput

One plate takes about 5 1/2 h to set up and analyze. For two or more plates to be run in one day, plate setup must be timed (staggered) to accommodate all the required steps. If two plates of the same assay (that can share prepared standards and reagents) are loaded 15 min apart, total set up and running time is ~6 h. To perform three different assays in a single day, assays must be started one hour apart and takes a total of about 8 h. These times do not include the assay preparation steps (i.e., tube labeling, CSF sample, and standard curve and working capture reagent preparation). We suggest running one assay with a few samples and QC samples at least ten separate times to gain proficiency and confidence.

10.7 Summary

We have provided a follow-up to our gene expression discovery approach for brain biomarkers for AD and presented four of the promising assays for CSF (three analytes; Ng, SNAP-25, VILIP-1) that arose from this work. We presented both of our Ng assays since the greater sensitivity of the rabbit assay could be useful in research involving smaller animals such as mice or rats where the CSF volume is very low. Assays for blood samples are also possible but not presented here. Many of the procedures for the assays reported here can be utilized with other assay platforms and we hope it will be instructive. As noted above, while the antibodies described here are currently not yet commercially available, discussion concerning their use can be made with Washington University Office of Technology Management or with the corresponding authors. Until a cure for AD is found, it is not known when these and other biomarkers will leave the research laboratory and enter the clinical laboratory. We remain optimistic that a cure will come soon.

Acknowledgments

The authors dedicate this chapter to our colleague, Dan L. Crimmins, Ph.D., whose passing on November 16, 2016, was a professional and personal loss to us. Dan was an outstanding protein chemist and friend, and his influence in the field of neuroscience is greatly missed. We acknowledge the many colleagues who have worked with us over the years, including: Kaj Blennow, David Brody, Mark Cervinski, Gregory Day, Mari DeMarco, Joe Gaut, David Holtzman, Eric Klawiter, Paul Kotzbauer, Omar Laterza, Jin-Moo Lee, Larry Lewis, Tina Lockwood, Amit Mathur, Tim Miller, Vijay Modur, John Morris, Suzanne Schindler, Rawan Tarawneh, Gregory Zipfel, and many others.

Disclosures: *JHL is named as a co-inventor on patents filed and managed by Washington University concerning brain biomarkers.*

References

1. Ladenson JH (2012) Reflections on the evolution of cardiac biomarkers. Clin Chem 58:21–24
2. Laterza OF, Modur VR, Ladenson JH (2008) Biomarkers of tissue injury. Biomark Med 2:81–92
3. Laterza OF, Modur VR, Crimmins DL et al (2006) Identification of novel brain biomarkers. Clin Chem 52:1713–1721
4. Lee J-M, Blennow K, Andreasen N et al (2008) The brain injury biomarker VLP-1 is increased in the cerebrospinal fluid of Alzheimer disease patients. Clin Chem 54:1617–1623
5. Tarawneh R, D'Angelo G, Macy E et al (2011) Visinin-like protein-1: diagnostic and prognostic biomarker in Alzheimer disease. Ann Neurol 70:274–285
6. Tarawneh R, Lee J-M, Ladenson JH et al (2012) CSF VILIP-1 predicts rates of cognitive decline in early Alzheimer disease. Neurology 78:709–719
7. Tarawneh R, Head D, Allison S et al (2015) Cerebrospinal fluid markers of Neurodegeneration and rates of brain atrophy in early Alzheimer disease. JAMA Neurol 72:656–665

8. Tarawneh R, D'Angelo G, Crimmins D et al (2016) Diagnostic and prognostic utility of the synaptic marker Neurogranin in Alzheimer disease. JAMA Neurol 73:561–571
9. Fagan AM, Xiong C, Jasielec MS et al (2014) Longitudinal change in CSF biomarkers in autosomal-dominant Alzheimer's disease. Sci Transl Med 6:226ra30
10. Kester MI, Teunissen CE, Sutphen C et al (2015) Cerebrospinal fluid VILIP-1 and YKL-40, candidate biomarkers to diagnose, predict and monitor Alzheimer's disease in a memory clinic cohort. Alzheimers Res Ther 7 (59)
11. Sutphen CL, Jasielec MS, Shah AR et al (2015) Longitudinal cerebrospinal fluid biomarker changes in preclinical Alzheimer disease during middle age. JAMA Neurol 72:1029–1042
12. Shahim P, Mattsson N, Macy EM et al (2015) Serum visinin-like protein-1 in concussed professional ice hockey players. Brain Inj 29:872–876
13. Ju Y-ES, Finn MB, Sutphen CL et al (2016) Obstructive sleep apnea decreases central nervous system-derived proteins in the cerebrospinal fluid. Ann Neurol 80:154–159
14. Crimmins DL, Herries EM, Ohlendorf MF et al (2017) Double monoclonal immunoassay for quantifying human Visinin-like Protein-1 in CSF. Clin Chem 63:603–604
15. Sutphen CL, McCue L, Herries EM et al (2018) Longitudinal decreases in multiple cerebrospinal fluid biomarkers of neuronal injury in symptomatic late onset Alzheimer's disease. Alzheimers Dement J Alzheimers Assoc 14:869–879
16. Schindler SE, Li Y, Todd KW et al (2019) Emerging cerebrospinal fluid biomarkers in autosomal dominant Alzheimer disease. Alzheimers Dement 15(5):655–665
17. Crimmins DL, Brada NA, Lockwood CM et al (2010) ETRAP (efficient trapping and purification) of target protein polyclonal antibodies from GST-protein immune sera. Biotechnol Appl Biochem 57:127–138
18. Todd J, Freese B, Lu A et al (2007) Ultrasensitive flow-based immunoassays using single-molecule counting. Clin Chem 53:1990–1995
19. Rissin DM, Kan CW, Campbell TG et al (2010) Single-molecule enzyme-linked immunosorbent assay detects serum proteins at subfemtomolar concentrations. Nat Biotechnol 28:595–599
20. Castaño EM, Maarouf CL, Wu T et al (2013) Alzheimer disease periventricular white matter lesions exhibit specific proteomic profile alterations. Neurochem Int 62:145–156
21. Kester MI, Teunissen CE, Crimmins DL et al (2015) Neurogranin as a cerebrospinal fluid biomarker for synaptic loss in symptomatic Alzheimer disease. JAMA Neurol 72:1275–1280
22. Remnestål J, Just D, Mitsios N et al (2016) CSF profiling of the human brain-enriched proteome reveals associations of neuromodulin and neurogranin to Alzheimer's disease. Proteomics Clin Appl 10:1275–1280
23. Willemse EAJ, De Vos A, Herries EM et al (2018) Neurogranin as cerebrospinal fluid biomarker for Alzheimer disease: an assay comparison study. Clin Chem 64:927–937
24. Davidsson P, Blennow K (1998) Neurochemical dissection of synaptic pathology in Alzheimer's disease. Int Psychogeriatr 10:11–23
25. Kvartsberg H, Duits FH, Ingelsson M et al (2015) Cerebrospinal fluid levels of the synaptic protein neurogranin correlates with cognitive decline in prodromal Alzheimer's disease. Alzheimers Dement J Alzheimers Assoc 11:1180–1190
26. Kvartsberg H, Portelius E, Andreasson U et al (2015) Characterization of the postsynaptic protein neurogranin in paired cerebrospinal fluid and plasma samples from Alzheimer's disease patients and healthy controls. Alzheimers Res Ther 7(40)
27. De Vos A, Jacobs D, Struyfs H et al (2015) C-terminal neurogranin is increased in cerebrospinal fluid but unchanged in plasma in Alzheimer's disease. Alzheimers Dement J Alzheimers Assoc 11:1461–1469
28. De Vos A, Struyfs H, Jacobs D et al (2016) The cerebrospinal fluid Neurogranin/BACE1 ratio is a potential correlate of cognitive decline in Alzheimer's disease. J Alzheimers Dis JAD 53:1523–1538
29. Kvartsberg H, Lashley T, Murray CE et al (2019) The intact postsynaptic protein neurogranin is reduced in brain tissue from patients with familial and sporadic Alzheimer's disease. Acta Neuropathol (Berl) 137:89–102
30. Brinkmalm A, Brinkmalm G, Honer WG et al (2014) SNAP-25 is a promising novel cerebrospinal fluid biomarker for synapse degeneration in Alzheimer's disease. Mol Neurodegener 9:53
31. Liu Y, Lv K, Li Z et al (2012) PACSIN1, a tau-interacting protein, regulates axonal elongation and branching by facilitating microtubule instability. J Biol Chem 287:39911–39924
32. Foerch C, Niessner M, Back T et al (2012) Diagnostic accuracy of plasma glial fibrillary acidic protein for differentiating intracerebral hemorrhage and cerebral ischemia in patients

with symptoms of acute stroke. Clin Chem 58:237–245

33. Mroczko B, Groblewska M, Zboch M et al (2015) Evaluation of visinin-like protein 1 concentrations in the cerebrospinal fluid of patients with mild cognitive impairment as a dynamic biomarker of Alzheimer's disease. J Alzheimers Dis JAD 43:1031–1037
34. Andreasson U, Perret-Liaudet A, van Waalwijk van Doorn LJC et al (2015) A practical guide to immunoassay method validation. Front Neurol 6:179
35. Westgard JO, Barry PL, Hunt MR et al (1981) A multi-rule Shewhart chart for quality control in clinical chemistry. Clin Chem 27:493–501
36. Bioporto biotinylation of antibodies. http://www.bioporto.com/Files/Images/Fra%20Files/Marketing-material/gRAD-Biotinylation-protocol_v01.pdf
37. Lago-Cachón D (2013) How much biotin could be in a biotinylated antibody? https://www.researchgate.net/post/How_much_biotin_could_be_in_a_biotinylated_antibody
38. Wu J (2000) Quantitative immunoassay: a practical guide for assay establishment, troubleshooting and clinical application. American Association for Clinical Chemistry

Chapter 13

Quantification of the Neurofilament Light Chain Protein by Single Molecule Array (Simoa) Assay

Christian Barro, Sarah Storz, Svenya Gröbke, Zuzanna Michalak, and Jens Kuhle

Abstract

Neurofilaments are cytoskeletal proteins specific for neurons. They are released upon neuronal injury and are increasingly recognized as a biomarker of disease activity in various neurological diseases. The development of a bead-based single molecule array (Simoa) assay for the neurofilament light chain provided a substantial improvement in sensitivity compared with conventional ELISA or electrochemiluminescence-based assays allowing reliable quantification in blood samples.

Key words Neurofilament light chain, Simoa, Biomarker

1 Introduction

Neurofilament light (NfL), medium (NfM), and heavy (NfH) chain are together with α-internexin and peripherin referred as the neuronal type IV intermediate filament (IF) proteins [1]. Neurofilaments (NF) are exclusively present in the cytoplasm of neuronal cells both in the central (CNS) and peripheral nervous system (PNS) [2]. Neuronal damage results in NfL release in the cerebrospinal fluid (CSF) and venous blood [3]. The levels of NfL, in both CSF and blood, have recently been studied as biomarkers for diagnosis and/or disease activity and progression in various diseases such as multiple sclerosis [4–8], ALS [9–12], Guillain–Barré syndrome or Alzheimer's disease [13]. Highly sensitive assays are necessary for reliable measurements in blood samples since NfL levels in serum and plasma are approximately 30–70-fold lower than those in CSF. Immunoblot and enzyme-linked immunosorbent assays are limited in sensitivity, and only more sensitive electrochemiluminescence and single molecule array (Simoa) assays recently enabled the detection of NfL in the easily accessible

Charlotte E. Teunissen and Henrik Zetterberg (eds.), *Cerebrospinal Fluid Biomarkers*, Neuromethods, vol. 168,
https://doi.org/10.1007/978-1-0716-1319-1_13,

blood compartment [6, 13–15]. We describe a highly sensitive Simoa-based assay for the quantification of NfL in blood and cerebrospinal fluid [6] (**Notes 1–4**).

2 Materials

The assay is run on an SIMOA HD-1 Analyzer platform (Quanterix, Lexington, United States) based on the Simoa technology [16, 17]. All reagents are stored at room temperature if not otherwise indicated.

2.1 Preparation of Calibrators

1. Bovine-lyophilized NfL (UmanDiagnostics, Umeå, Sweden; −80 °C).
2. Tween 20 (Sigma Aldrich, #P9416).
3. Tris-buffered saline (TBS): 100 mM tris base and 1.5M NaCl (pH 7.5).
4. Bovine nonfat milk powder (e.g., Migros, Switzerland).
5. Heteroblock (Omega Biologicals, Montana, United States; −20 °C, up to 6-month storage +4 °C).

2.2 Coating of Beads with Capture Antibody

1. Anti-NfL capture antibody (mouse monoclonal, clone 47:3, UD1), UmanDiagnostics; −20 °C, up to one-month storage +4 °C).
2. "Simoa Homebrew Assay Development Kit" (Quanterix, #101354) (+4 °C):
   - singleplex carboxylated paramagnetic beads.
   - Bead Conjugation Buffer
   - Bead Wash Buffer
   - Bead Diluent
   - Bead Blocking Buffer
3. Ultra-0.5 Centrifugal Filter Unit with Ultracel-50 membrane (Amicon, #UFC505096).
4. 1-ethyl-3-(3-dimethylaminopropyl) carbodiimide hydrochloride (EDC, ThermoFisher Scientific, #77149; −20 °C).
5. Rotator (HulaMixer, Thermofisher Scientific, #15920D).
6. Single Tube Magnetic Stand (magnetic separator) for magnetic bead concentration (Ambion, #AM10055).
7. Spectrophotometer (280 nM, e.g., Nanodrop 2000, Thermofisher Scientific, #ND-2000).

2.3 Measuring Protocol

1. Paramagnetic beads coated with anti-NfL capture antibody (clone 47:3, UD1, UmanDiagnostics, +4 °C).

2. Anti-NfL biotinylated detector antibody (mouse monoclonal, clone 2:1, UD3, UmanDiagnostics; −20 °C).
3. Tween 20.
4. TBS.
5. Bovine nonfat milk powder.
6. Heteroblock.
7. "Simoa Enzyme & Substrate Kit" (Quanterix, #101361, +4 °C):
   - Streptavidin beta galactosidase (SBG) concentrate
   - SBG diluent
   - Resorufin β-D-galactopyranoside (RGP)
8. Pierced plate seals (Excel scientific, #EZP-100)
9. "Simoa Discs Kit" (Quanterix, #100227):
   - Conical tubes
   - Discs
   - Cuvettes
   - 96-well plates
   - Simoa pipettor tips
10. "Simoa buffer 1&2 Combo pack" (Quanterix, #100488):
    - System Buffer 1
    - System Buffer 2
11. Simoa Sealing Oil (Quanterix, #100206).

3 Method

3.1 Preparation of Calibrators

Calibrators are produced by serial dilution of calibrator stock concentrate in TBS, pH 7.5, containing 1% bovine nonfat milk powder, 0.1% Tween 20, 300 μg/mL Heteroblock at following concentrations: 10000 (only for CSF measurements), 2000, 400, 80, 16, 3.2, 0.64, 0.32, 0 pg/mL. Multiple sets of calibrators are batch prepared and stored (270 μl per tube, −80 °C).

3.2 Coating of Beads with Capture Antibody (Note 5)

Move buffers and reagents on ice; UD1 should be room temperature. All incubations are carried out at room temperature on rotator (Hulamixer: orbital 5–5; reciprocal 90–5; vibration 5–5).

(a) *Buffer exchange 78 μg of UD1 to bead conjugation buffer using Ultra-0.5 centrifugal filter:*

1. Pipette the required volume of UD1 to an Ultra-0.5 Centrifugal Filter.

2. Add Bead Conjugation Buffer to Ultra-0.5 Centrifugal Filter to a total volume of 500 μL. Invert tube three times to mix after each addition of Bead Conjugation Buffer.
3. Centrifuge at 14000 $\times g$, 5 min, discard flow-through.
4. Add 450 μL Bead Conjugation Buffer (be quick).
5. Centrifuge at 14000 $\times g$, 5 min, discard flow-through.

   Repeat **steps 4** and **5** overall two times.
6. Place the Ultra-0.5 Centrifugal Filter upside down into a new collection tube, spin at 1000 $\times g$ for 2 min to collect concentrated and washed UD1.
7. Remove the filter from the collection tube and wash both membranes by pipetting 50 μL of Bead Conjugation Buffer against the filter membranes to recover additional UD1, aspirate and pipette against each side of the filter three times.
8. Carefully place back inverted and washed filter into the same collection tube as **step 6** spin at 1000 $\times g$ for 2 min to collect concentrated and washed UD1.
9. Measure the concentration of UD1 (wavelength 280 nm, for example, in Nanodrop program "A280", use Bead Conjugation Buffer as blank).
10. Measure volume of recovered UD1 and dilute to 0.3 μg/μL with Bead Conjugation Buffer (this is our **final volume of capture antibody**).
11. Briefly vortex antibody and store on ice.

(b) *Buffer exchange 1.4 × 10⁶ beads per μL of recovered antibody to bead wash buffer and finally to bead conjugation buffer:*

1. Transfer the required volume of beads to a 1.7 mL Eppendorf tube:

$$\text{Required volume of beads} = \frac{1.4 \times 10^6 \left(\frac{\text{beads}}{\mu L}\right) * \text{final volume of capture antibody } (\mu L)}{\text{Stock concentration } \left(\frac{\text{beads}}{\mu L}\right)}$$

2. Buffer Exchange the Beads to Bead Wash Buffer

   Perform wash steps with the same volume as "final volume of capture antibody" (*see* **step A.10**), abbreviated as "x" in this protocol.

   (a) Pulse spin tube, place it in a magnetic separator (1 min), aspirate the liquid while the beads are attracted by the magnet, remove the tube from magnet.

(b) Add x μL Bead Wash Buffer, carefully resuspend beads with the pipette and vortex for 5 s.

(c) Pulse spin tube, place it in a magnetic separator (1 min), aspirate the liquid, remove the tube from magnet.

Repeat 2.1–2.3 overall three times.

3. Buffer exchange the beads to Bead Conjugation Buffer.

(a) Add x μL Bead Conjugation Buffer, carefully resuspend beads with the pipette and vortex 5 s.

(b) Pulse spin tube, place it in a magnetic separator (1 min), aspirate the liquid, remove the tube from magnet.

(c) Repeat 3.1 and 3.2 overall two times.

(d) Ensure all liquid is removed from the last wash step.

(e) Add a volume of Bead Conjugation Buffer equivalent to 95% of the final volume of UD1 (*see* **step A.10**), carefully resuspend beads and vortex 5 s.

(f) Pulse spin tube and store on ice until use.

(c) *Activate the Beads*

1. Carefully add 1 mL ice cold Bead Conjugation Buffer to 10 mg vial of EDC (**Note 6**).
2. Vortex and invert EDC until dissolved.
3. Add ice cold EDC to the cold washed beads: EDC volume = 5% of the final volume of capture antibody (see **step A.10**); vortex 10 s.
4. Mix on rotator for 30 min.

(d) *Conjugate UD1 with the activated beads*

1. Pulse spin tube with washed beads, place it in magnetic separator (1 min), aspirate the liquid, remove the tube from magnet (be quick).
2. Add x μL of ice cold Bead Conjugation Buffer, vortex 5 s.
3. Pulse spin tube, place it in a magnetic separator (15 s), aspirate the liquid, remove the tube from magnet (be quick).
4. Add complete volume of 0.3 μg/μL UD1 (ice cold), vortex 10 s.
5. Mix on rotator for 2 h.

(e) *Blocking step for coated beads*

1. Pulse spin tube, place it in magnetic separator (1 min), and transfer supernatant to a new tube (keep as "supernatant").
2. Remove tube from magnet, add x μL Bead Wash Buffer, vortex 5 s.

3. Pulse spin tube, place it in magnetic separator (1 min), and transfer supernatant to a fresh tube (keep as "wash 1").
4. Remove tube from magnet, add x μL Bead Wash Buffer, vortex 5 s.
5. Pulse spin tube, place it in magnetic separator (1 min), aspirate the liquid.
6. Remove tube from magnet add x μL Bead Blocking Buffer, vortex 5 s.
7. Mix on rotator for 30 min.

(f) *Final step*

1. Pulse spin tube, place it in magnetic separator (1 min), aspirate the liquid.
2. Remove tube from magnet, add x μL Bead Wash Buffer, vortex 5 s.
3. Repeat **steps 1** and **2** overall twice.
4. Pulse spin tube, place on magnetic separator (1 min), aspirate the liquid.
5. Remove tube from magnet, add a volume of Bead diluent (same volume as the final volume of capture antibody, **step A.10**), vortex 5 s.
6. Pulse spin tube and store coated and blocked beads at 4 °C until use.

(g) *Characterization*

In order to quantify the efficiency of the coupling reaction, measure UD1 concentration in "supernatant" and "wash 1" by spectrophotometer (wavelength: 280 nm, e.g., Nanodrop program A280; blank for "supernatant": Bead Conjugation Buffer; blank for "wash 1": Bead Wash Buffer).

$$\text{Efficiency} = \frac{\text{Initial mass of Ab} - (\text{mass of Ab in supernatant} + \text{mass of Ab in Wash 1})}{\text{Initial mass of Ab}}$$

The final estimated bead concentration (coated beads) is 1.26×10^6 beads/μL (assuming a 10% loss during the procedure).

3.3 Measuring Protocol

Serum (and EDTA plasma) is diluted on bench 1:4 and CSF 1:10. Below calculations for buffer preparations are for 1 plate with 40 serum samples and 8 calibrators, each measured in duplicate from a single well.

(a) *Diluents preparation*

1. Sample diluent (TBS, 0.1% Tween, 1% bovine nonfat milk powder, 400 μg/mL Heteroblock for serum or 333 μg/mL for CSF): 180 μL for serum or 216 μL for CSF samples. For 1 plate with 40 serum samples to be measured: 40 × 180 μL = 7200 μL of sample diluent (**Note 7**).
2. Antibody diluent (TBS, 0.1% Tween, 1% bovine nonfat milk powder, 300 μg/mL Heteroblock):
   130 μL per sample and calibrator and an additional 600 μL for each conical tube used for beads and detector antibody dilution. For 1 plate: 48 × 130 μL + 600 μL + 600 μL = 7440 μL of antibody diluent (**Note 7**).
3. SBG diluent is ready to use.

(b) *Plate preparation*

1. Calibrators are measured neat, serum, and CSF samples in this version of the protocol are diluted in a quanterix 96-well plates 1:4 and 1:10, respectively (final volume in well: 240 μL) (please check Quanterix protocols for automated dilution protocols).
2. Prepare samples and calibrators before pipetting:
   vortex samples and calibrators.
   centrifuge samples (10 min, 4400 ×*g*; not calibrators).
3. Pipette 240 μL of each calibrator in each calibrator well.
4. Pipette 180 μL of sample diluent for serum or 216 μL for CSF in each sample well.
5. Pipette 60 μL of serum or 24 μL of CSF sample in each sample well.
6. When pipetting each sample, mix by pipetting up and down four times.
7. A pierced plate seal is then applied to the plate to limit evaporation.

(c) *Reagents preparation*

1. SBG 150pM is obtained by diluting SBG concentrate in SBG diluent: 200 μL per sample and calibrator, an additional 600 μL as dead volume for the conical tubes used.
2. Anti-NfL biotinylated detector antibody is diluted in antibody diluent to 0.1 μg/mL: 60 μL per sample and calibrator and additional 600 μL as dead volume for the conical tubes used.
3. The stock of coated NfL beads (1.26×10^6 beads/μL) is removed from the storage buffer, washed with antibody

diluent, and diluted to a final concentration of 20.16×10^3 beads/μL: 70 μL per sample and calibrator and an additional 600 μL as dead volume for the conical tube.

(a) Pipette the required amount of coated NfL beads in a new tube.
(b) Pulse spin tube.
(c) 1-min incubation on a magnetic separator.
(d) Remove storage buffer.
(e) Remove the tube from the magnet and resuspend beads in 200 μL of antibody diluent.
(f) Pulse spin tube.
(g) 1-min incubation on a magnetic separator.
(h) Remove antibody diluent.
(i) Remove the tube from the magnet, take 200 μL from the final volume prepared for beads dilution in the conical tube and resuspend the beads.
(j) Add the resuspended beads in the conical tube.

(d) *Machine assay definitions and procedure:*

Use a two-step Assay Neat 2.0 protocol:

Step 1

(a) 100 μL of calibrator/diluted sample are pipetted in the cuvette.
(b) 25 μL of beads are added in the cuvette.
(c) 20 μL of detector are added in the cuvette.
   - Incubation step: 47 cadences (1 cadence = 45 s).
   - Wash step

Step 2

(a) 100 μL of SBG are added in the cuvette.
   - Incubation step: 7 cadences
   - Wash step
   - 25 μL of RGP are added
   - 15 μL of the solution are loaded on a single array composed of approximately 240,000 microwells (46 fL), part of a 24-array disc
   - Reading step

4 Notes

4.1 Note 1. Intra- and Inter-Assay Precision

Intra- and inter-assay precision were determined by duplicate measurements of three native serum and three native CSF samples in 22 runs for serum and 12 runs for CSF on separate days. The mean intra-assay precision in the three serum samples was: 5.6% (13.3 pg/mL, sample 1), 6.9% (22.5 pg/mL, sample 2), 5.3% (236.5 pg/mL, sample 3); in the three CSF samples: 2.5% (572.6 pg/mL, sample 4), 0.7% (1601.8 pg/mL, sample 5), 3.8% (6110.2 pg/mL, sample 6). The mean inter-assay precision was determined as coefficient of variation (CV) between the mean concentration of the samples in each run. In serum, the mean inter-assay precision was: 11.3% (sample 1), 9.3% (sample 2), 6.4% (sample 3); in CSF: 10.1% (sample 4), 6.2% (sample 5), 15.5% (sample 6) [6].

4.2 Note 2. Analytical Sensitivity

Sensitivity was estimated as the concentration of the lowest calibrator fulfilling acceptance criteria: CV of concentration of duplicate determination below 20% and recovery 80–120% [18]. The sensitivity reached was of 0.32 pg/mL (taking dilution of samples into account, sensitivity was 1.28 pg/mL for serum, and 3.2 pg/mL for CSF).

4.3 Note 3. Parallelism and Selectivity

Parallelism was examined in four native serum and CSF samples diluted 1:2, 1:4, 1:8, 1:16. All sample concentrations corrected for dilution factor were within 30% of expected range (Fig. 13.1). Selectivity was determined in four serum and CSF native samples spiked with 5, 50, 200 pg/mL, and 500, 2000 pg/mL, respectively. The mean recovery in serum for spiked concentration was: (a) 5 pg/mL: 109%, (b) 50 pg/mL: 119%, (c) 200 pg/mL: 94%; in CSF: (a) 500 pg/mL: 125%, (b) 2000 pg/mL: 124%.

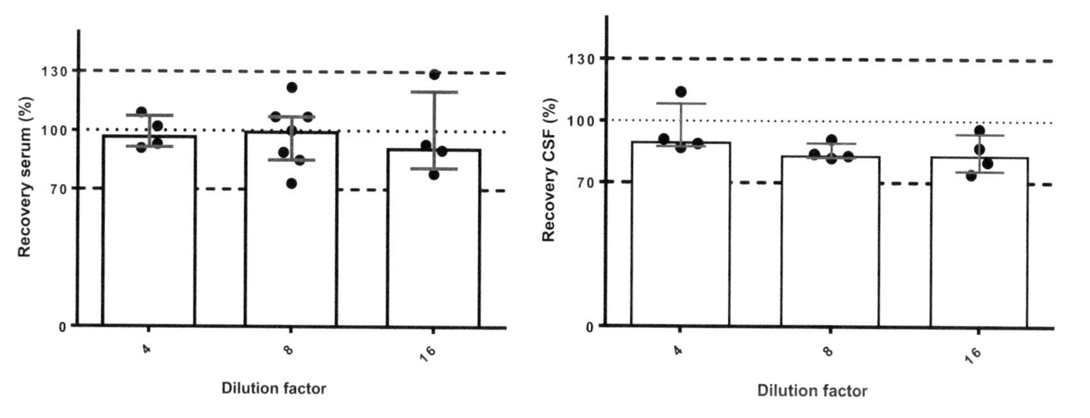

Fig. 13.1 Parallelism. Percentage change from the reference dilution factor 2 (100%). Serum on the left, and CSF on the right was serially diluted 1:4, 1:8, 1:16. The columns represent the median values, the red bars the interquartile range. Validity range 70–130% (dotted bold lines)

4.4 Note 4. Assay Stability over Time

Six samples were measured at baseline and after 6 months with different antibody lot and different operator: (a) 21.1 vs 21.0 pg/mL, CV: 0.3%; (b) 338.0 vs 365.1 pg/mL, CV: 5.4%; (c) 33.0 vs 32.7 pg/mL, CV: 0.7%; (d) 15.9 vs 18.0 pg/mL, CV: 8.8%; (e) 48.7 vs 38.9 pg/mL, CV: 15.8%; (f) 176.8 vs 213.2 pg/mL, CV: 13.2%. The mean CV was 8.5%.

4.5 Note 5. Reducing Variability within a Project

It is recommended to produce enough beads sufficient for an entire project to minimize potential batch-to-batch effects caused by beads coupling procedure.

4.6 Note 6. Use of EDC

Remove EDC from the freezer just prior to use. Once reconstituted EDC has a concentration of 10 mg/mL. The working concentration of EDC is 0.5 mg/mL: volume of EDC solution: 95% Bead Conjugation Buffer and 5% EDC (e.g., antibody final volume: 200 μL: 95% of 200 μL = 190 μL of Bead Conjugation Buffer and 5% of EDC = 10 μL). Dissolve EDC just before adding it to the beads.

4.7 Note 7. Diluents Preparation

Prepare the diluents in 15 mL or 50 mL falcon tubes and as safety consider producing some more diluent than needed, for example, additional 2 mL.

References

1. Perrot R, Berges R, Bocquet A, Eyer J (2008) Review of the multiple aspects of neurofilament functions, and their possible contribution to neurodegeneration. Mol Neurobiol 38 (1):27–65
2. Godsel LM, Hobbs RP, Green KJ (2008) Intermediate filament assembly: dynamics to disease. Trends Cell Biol 18(1):28–37
3. Reiber H (2003) Proteins in cerebrospinal fluid and blood: barriers, CSF flow rate and source-related dynamics. Restor Neurol Neurosci 21 (3–4):79–96
4. Teunissen CE, Dijkstra C, Polman C (2005) Biological markers in CSF and blood for axonal degeneration in multiple sclerosis. Lancet Neurol 4(1):32–41
5. Disanto G, Adiutori R, Dobson R, Martinelli V, Dalla CG, Runia T et al (2016) Serum neurofilament light chain levels are increased in patients with a clinically isolated syndrome. J Neurol Neurosurg Psychiatry 87 (2):126–129
6. Disanto G, Barro C, Benkert P, Naegelin Y, Schadelin S, Giardiello A et al (2017) Serum neurofilament light: a biomarker of neuronal damage in multiple sclerosis. Ann Neurol 81 (6):857–870
7. Piehl F, Kockum I, Khademi M, Blennow K, Lycke J, Zetterberg H et al (2018) Plasma neurofilament light chain levels in patients with MS switching from injectable therapies to fingolimod. Mult Scler 24:1046–1054
8. Novakova L, Zetterberg H, Sundstrom P, Axelsson M, Khademi M, Gunnarsson M et al (2017) Monitoring disease activity in multiple sclerosis using serum neurofilament light protein. Neurology 89:2230
9. Lu CH, Macdonald-Wallis C, Gray E, Pearce N, Petzold A, Norgren N et al (2015) Neurofilament light chain: a prognostic biomarker in amyotrophic lateral sclerosis. Neurology 84(22):2247–2257
10. Menke RA, Gray E, Lu CH, Kuhle J, Talbot K, Malaspina A et al (2015) CSF neurofilament light chain reflects corticospinal tract degeneration in ALS. Ann Clin Transl Neurol 2 (7):748–755
11. Feneberg E, Oeckl P, Steinacker P, Verde F, Barro C, Van Damme P et al (2018) Multicenter evaluation of neurofilaments in early

symptom onset amyotrophic lateral sclerosis. Neurology 90(1):e22–e30

12. Weydt P, Oeckl P, Huss A, Muller K, Volk AE, Kuhle J et al (2016) Neurofilament levels as biomarkers in asymptomatic and symptomatic familial amyotrophic lateral sclerosis. Ann Neurol 79(1):152–158
13. Gaiottino J, Norgren N, Dobson R, Topping J, Nissim A, Malaspina A et al (2013) Increased neurofilament light chain blood levels in neurodegenerative neurological diseases. PLoS One 8(9):e75091
14. Kuhle J, Barro C, Andreasson U, Derfuss T, Lindberg R, Sandelius A et al (2016) Comparison of three analytical platforms for quantification of the neurofilament light chain in blood samples: ELISA, electrochemiluminescence immunoassay and Simoa. Clin Chem Lab Med 54(10):1655–1661
15. Gisslen M, Price RW, Andreasson U, Norgren N, Nilsson S, Hagberg L et al (2016) Plasma concentration of the Neurofilament light protein (NFL) is a biomarker of CNS injury in HIV infection: a cross-sectional study. EBiomedicine 3:135–140
16. Rissin DM, Kan CW, Campbell TG, Howes SC, Fournier DR, Song L et al (2010) Single-molecule enzyme-linked immunosorbent assay detects serum proteins at subfemtomolar concentrations. Nat Biotechnol 28(6):595–599
17. Wilson DH, Rissin DM, Kan CW, Fournier DR, Piech T, Campbell TG et al (2016) The Simoa HD-1 analyzer: a novel fully automated digital immunoassay analyzer with single-molecule sensitivity and multiplexing. J Lab Autom 21(4):533–547
18. Valentin MA, Ma S, Zhao A, Legay F, Avrameas A (2011) Validation of immunoassay for protein biomarkers: bioanalytical study plan implementation to support pre-clinical and clinical studies. J Pharm Biomed Anal 55(5):869–877

Index

Charlotte E Teunissen and Henrik Zetterberg (eds.), *Cerebrospinal Fluid Biomarkers*, Neuromethods, vol. 168, https://doi.org/10.1007/978-1-0716-1319-1,

T

V

Printed in the United States
by Baker & Taylor Publisher Services